BIM and Construction Management

BIM and Construction Management

Proven Tools, Methods, and Workflows

Second Edition

Brad Hardin

Dave McCool

WILEY

Acquisitions Editor: MARIANN BARSOLO
Development Editor: TOM CIRTIN
Technical Editor: JANA CONOVER
Production Editor: REBECCA ANDERSON
Copy Editor: ELIZABETH WELCH
Editorial Manager: MARY BETH WAKEFIELD
Production Manager: KATHLEEN WISOR
Associate Publisher: JIM MINATEL
Book Designer: FRANZ BAUMHACKL
Proofreader: NANCY CARRASCO
Indexer: J & J INDEXING
Project Coordinator, Cover: NICK WEHRKAMP
Cover Designer: WILEY
Cover Image: © PAOLO GAETANO ROCCO / iSTOCKPHOTO

Copyright © 2015 by John Wiley & Sons, Inc., Indianapolis, Indiana

Published simultaneously in Canada

ISBN: 978-1-118-94276-5
ISBN: 978-1-118-94278-9 (ebk.)
ISBN: 978-1-118-94277-2 (ebk.)

For general information on our other products and services or to obtain technical support, please contact our Customer Care Department within the U.S. at (877) 762-2974, outside the U.S. at (317) 572-3993 or fax (317) 572-4002.

Wiley publishes in a variety of print and electronic formats and by print-on-demand. Some material included with standard print versions of this book may not be included in e-books or in print-on-demand. If this book refers to media such as a CD or DVD that is not included in the version you purchased, you may download this material at http://booksupport.wiley.com. For more information about Wiley products, visit www.wiley.com.

Library of Congress Control Number: 2015930545

SKY10061409_113023

For my parents, who let me draw on the walls. For my great kids who are loved by their geek dad and for my beautiful wife who is beyond supportive.

— B.H.

For Paul Vance, my high school technical drawing teacher at Vestavia Hills, who found and fostered a passion that has shaped my career.

— D.M.

Acknowledgments

I would like to thank my wife, Iris; daughter, Lucia; and son, Wesley for supporting the late nights and shared time of weekends with this project. I couldn't have done it without your support. I'm blessed to have family, Jen, Dave, Mom, and Dad and friends, Joe Moerke, Eric Glatzl, and DJ who helped as much as they have. Lulu, I have no more pages left in my "chapter book" to do…

I would also like to thank my co-author, Dave McCool, who agreed to partner up and take this book project head on. Dave contributed great insights and valuable content and supported many good discussions on what BIM "really is" and how best to tell that story. It has been a pleasure working with such an industry leader.

I'm thankful to the firms, colleagues, industry organizations, and academics who let us use their work, insights, and images for case studies. Thanks to Black & Veatch for allowing me the time to see this work completed. I'm hopeful the design and construction industry takes this content and uses it to accelerate positive change in this industry I am so passionate about.

– Brad Hardin

First off, I want to thank Brad Hardin for this amazing opportunity. He's been a great friend and mentor throughout this journey, and I'm excited about our next adventure. I also want to thank my dad (Jim McCool, PE, CEM, CxA, LEED AP), who has been a role model father and mentor. Dad, you're not allowed to get any more acronyms. It won't fit on our business cards! My whole family has been incredibly supportive and encouraging throughout this whole process. Meg, thank you for the edits and brainstorming sessions. Mom, thank you for the counseling. Emily, thank you for waiting till I was done.

I would also like to thank the many others who have mentored and supported me over the years. This book wouldn't have happened without you: Tommy Duncan, Morgan Duncan, Bill Hitchcock, Dianne Gilmer, Trey Clegg, Mike Dunn, Mike Mitchell, Jason Lee, Sam Hardie, Sarah Carr, Derek Glanvill, Randy Highland, Chad Dorgan, Jim Mynott, Simon Peters, Shannon Lightfoot, Enrique Sarmiento, Connor Christian, John Grady, Brasfield & Gorrie, and the entire family at McCarthy Building Companies. To Dr. H and Dianne, thank you for taking a chance with a psychology major. I will forever be indebted to you and the Construction Engineering Management master's program you created at the University of Alabama at Birmingham (UAB).

From both of us, a very big thank you to the Wiley team: Pete Gaughan, who saw value in this project; Mariann Barsolo, for the patience and prodding in getting this done; Thomas Cirtin, Becca Anderson, Liz Welch, and Nancy Carrasco for helping us say what we meant to and making us sound smarter than we are; and Jana Conover, for taking on a new challenge and checking the technical components and tutorials.

– Dave McCool

About the Authors

Brad Hardin is the Chief Technology Officer for Black & Veatch, a global engineering and construction firm. He is a LEED-accredited architect, an ENR 20 under 40 recipient and is an advisory board member of the New School of Architecture. He has written numerous articles, given numerous presentations, and enjoys participating in industry events to further the cause of BIM, technology, and AEC startups in the design and construction market. He is a co-founder of Virtual Builders (www.virtualbuilders.com) the world's first nonprofit software- and association-agnostic certification and open source BIM intelligence development community for the design construction and operations industry. He lives in Kansas City with his wife, Iris; his two children, Wes and Lulu; and a dog named Shiloh.

Dave McCool is the Director of Virtual Design and Construction at McCarthy Building Companies. He holds a master's degree in engineering, DBIA, and LEED accredidation, but has realized that his BS in psychology is much more useful in the construction industry than any of the other credentials. He has lectured at multiple university and industry events, and has held chair positions for both AIA and NBIMS committees. He is also a co-founder of Virtual Builders. Originally from Alabama, he now lives in Los Angeles, where he enjoys the sunny weather, trying to surf, and playing music on the weekends.

Contents

Introduction

This book shares a rounded perspective of how BIM and enabling technologies are changing the way we collaborate and distribute information. As an industry, we are constantly facing new challenges in the field of construction. This book will show how many of these challenges are being addressed with cutting-edge tools, leveraged with experience, and a practical application of the "right tools for the right job." There is a shift happening in the construction management market in the context of technology, and this book serves as a catalyst for more fundamental changes that create positive outcomes.

The first version of *BIM and Construction Management: Proven Tools, Methods, and Workflows* (Sybex, 2009) by Brad Hardin was written just as the construction industry had largely begun to pay attention to this exciting new tool and process: building information modeling. Since then, the pace and transformational changes that have cascaded through the industry have been remarkable. Now clash detection, 4D sequencing, model estimates, and walk-throughs have become table stakes. Customers are now asking about Big Data, model to prefabrication, life-cycle energy modeling, project partnering approaches, and how BIM can mitigate other risk factors during construction. And still the pace of technology continues to move at an incredible rate.

The focus has now broadened from beyond BIM and the question is being asked, "If BIM can change the construction management business so significantly, what else can BIM do and what possibilities do other technologies hold?"

This broader questioning of the tools, teamed with economic challenges, has given rise to a technological renaissance in the construction community. Because of the recession, many firms were forced to refocus and question the best way to deliver construction product to customers under new margin and overhead constraints. The early successes of BIM gave many organizations a starting point to focus on. Some firms didn't stop at BIM and began taking a deeper look at not only the technology, but the underlying processes that were built around these tools. In this broader examination, there has been a significant push for innovation in construction technology and processes as well as enabling behaviors.

So What's Changed?

To begin, innovations in technology such as wearable tech, cloud-based collaboration, and the continued removal of hardware constraints have opened many doors for continued impact. Additionally, process innovations such as lean planning and an overall challenging of many of the traditional constructs of the construction industry, such as CPM scheduling, documentation strategies, contract arrangements, and the roles of design and construction teams at large have brought about a refreshing analytical perspective to the way we deliver work. The result has been an exciting view "into the looking glass" of what the future of our industry holds. We may very well be at the point of another paradigm shift in which the analysis of industry norms combined with more informed construction consumers could bring about the next revolution in the construction industry. These customers continue to be less willing to pay for our inefficiencies as an industry. Because of these factors, this movement will focus on *results-based deliverables*, with technology acting as a baseline expectation instead of an innovation to deliver on the "best value" promise.

Arguably, all industries are becoming increasingly reliant on technology to uncover previously unexplored value potential. The construction industry is no different. Almost daily, it seems that companies and individuals are coming up with an array of potential opportunities for improvement that will surely shape the way we do work for years to come. On average, there are 20,000 applications a month being uploaded to Apple's iOS store. Technologies like Google Glass, tablets, photogrammetry, mobile applications and a host of other potential hardware and software improvements are beginning to migrate into the way we do business; see the article at `http://readwrite.com/2013/01/07/apple-app-store-growing-by#awesm=~oDoS5C7qwveOnJ`. What impact will these tools have? How much safer will they make our jobsites? How do we quickly analyze the value of these tools at a pace that keeps up with the market? Questions like these led us to believe that the construction industry needed a more rounded take on not just BIM and how it relates to construction management, but an overall perspective of what these tools are and the enabling ecosystem that shows a more holistic approach to the way we can improve the design and construction industry.

> *You can't connect the dots looking forward; you can only connect them looking backwards. So you have to trust that the dots will somehow connect in your future....*

> – Steve Jobs

Because of this broadened focus, this new edition will look at the results desired and show the process of selecting tools to get there. This book will also look at some of the cutting-edge applications that either work in tandem with BIM or operate outside of it, and provide significant value to users during the construction process. Some of these tools may relate to each other, whereas some may not. However, it is important to highlight where information links to other tools and where the gaps are because they show the opportunities for improvement within our workflows as an industry.

An additional benefit of broadening the scope and context of this work is to better understand best practices on how construction management companies quickly analyze tools as they become available and how to implement the tools that create significant value and identify disruptive ones.

Trust is everything. And this book delves deeper into the enabling behaviors and mind-sets that make the use of BIM and technology successful. Significant research has been done on this topic and the better outcomes as a result of teams having the right behaviors as well as better understanding people's personalities and working dynamics. According to *Profitable Partnering for Lean Construction* (Wiley-Blackwell, 2004) by Clive Thomas Cain, "Strategic partnering can deliver significant savings, of up to 30% in the cost of construction." One of the major benefits in BIM is the unlocked potential that comes from having trusted information available early that make for better informed decisions. Similarly, understanding your project partner's abilities and the ways they work can make for a more meaningful dialogue and ultimately better workflows.

Lastly, this book will introduce the concept of information flow in construction management. While relatively new to the construction management space, flow is something that is critical in the performance of construction projects. If you have a project with good flow, teams distribute and receive information on time, in the desired format, and with clear expectations of the desired outcomes. Without good flow, projects jerk and start like a car without a consistent fuel supply, constantly grabbing at the next bit of information that will allow them to proceed with their tasks, all at the expense of the overall project as someone is consistently waiting on someone else. The goal of the Japanese term *Genjitsu* is the passing of reliable and accurate data to your fellow team members. The goal of BIM is to ultimately drive waste from the way we deliver construction projects to construction consumers. This book will show the value in information flow planning and how it is accomplished by focusing on passing the right information to project stakeholders rather than volumes of disconnected data.

Who Should Read This Book?

This book was written for those who wish to learn more about better ways to holistically leverage BIM and technology in the construction process. Those who will find this book useful may be:

- Designers wanting to better understand construction managers' tools and processes
- Construction managers looking to better understand the ways BIM and technology can be used to create better outcomes
- Subcontractors and project stakeholders looking to find ways to become a more valued player
- Owners and construction consumers who want to be more informed and who wish to create a more successful project and project team
- Students who want to grow their knowledge of BIM and technology in construction and learn how they should challenge the constructs of the industry where there are better ways of working

In particular, this book is for those who are interested in creating a better paradigm of delivering the built environment. It is not intended to be the sole definition of how to use BIM on a construction project, nor is it intended to be the definitive "how-to" guide. Rather this book is meant to delineate a way of looking for and delivering value in using BIM and technology. Readers will be shown how to challenge traditional deliverables and thinking, and how best to combine available project information and technology and pull these toward a desired end state.

How to Use This Book

This book is structured, in a linear fashion, similar to how a construction project would progress throughout the various stages until completion. The contents will walk users through tools that may be applied at various points along a project timeline and what the anticipated outcomes and results should be. The tools and processes highlighted are meant to be contextual and the concepts shown are for reference. To be sure, just as this book is printed, new tools are being introduced into the market that may very well improve on some of those mentioned. By reading the chapters in sequence, you should gain an understanding of how the tools can work through a construction project, what information is required, what the outputs are, and where that information may or may not connect to other systems.

This book will show how to establish agreed-upon metrics in the beginning of a project to gauge project success from which the team as a whole will be measured. We will show screenshots of various workflows and how some processes work to illustrate interfaces, information required, and level of effort. Lastly, case studies will be used on relevant topics to show real-world examples of the tools and processes in action to further explore the use case and context of the topics within the book.

The chapters in this book are as follows:

Chapter 1: Why Is Technology So Important to Construction Management   Chapter 1 has two purposes; the first is to act as a preview of the more detailed contents within the later chapters, as well as exploring where BIM and technology is being applied in construction management. This chapter will show ways BIM is used in construction as we collaborate together to virtually build structures and what impacts the various tools have in the BIM process. This chapter will cover at a high level the places where BIM and technology can provide additional value. These areas of focus include a linear approach to the project cycle. We will walk through topics such as team engagement, pursuit and marketing, preconstruction, construction, and closeout with many other detailed subpoints such as contracts, scheduling, logistics, and estimating to give further perspective.

Finally, this chapter will discuss industry trends relating to where technology and BIM is headed and show you how to get ahead of the technology curve. The chapter concludes with how to achieve leadership buy-in, strategies to attract and engage the right talent to drive the use of the tools, and the results the industry has seen.

Chapter 2: Project Planning   Chapter 2 includes a detailed overview and results-driven approach to how to set up your project to succeed. As it relates to BIM and technology, project planning is of critical importance to a construction project and is often a driver for a successful project. This chapter will walk you through standard contract delivery vehicles and the pros and cons of using technology in each. This chapter will also focus on defining the various uses of BIM and the resources required to execute them successfully. Lastly, it will focus on information flow, where project participants have a clear understanding of their role and responsibility in a project and aligned expectations throughout the entire project team. The chapter will identify current BIM contract language from industry organizations and explain how to create meaningful language derived from the BIM execution plans and checklists available in the market.

Chapter 3: How to Market BIM and Win the Project   How do you market your BIM and technology capabilities to customers and the industry? This chapter will walk readers through the process of how to show your capabilities, share results, and deliver focused solutions that are customized for each customer without having to constantly invest in new tools and technology. This chapter will explore with readers the dangers in overpromising on new technologies that haven't been proven and what impacts that can have downstream. Most important, it shows how to establish a trust-based technology delivery platform that will not only satisfy customers' needs but also drive future business opportunities as a mutual partnership.

Chapter 4: BIM and Preconstruction   Since the introduction of BIM into the construction management marketplace, preconstruction has been a key focus area for the use of the tools. Partly due to the nature of BIM and the ability to create and use information early as well as a means for better collaboration and exchange between project teams, BIM has grown in use and possibilities in the area of preconstruction. Chapter 4 explores how BIM and technology is being integrated throughout preconstruction activities such as scheduling, logistics, estimating, constructability analysis, visualization, and prefabrication planning.

Chapter 5: BIM and Construction   Chapter 5 is dedicated to BIM during construction. This chapter focuses on the nuts and bolts of using BIM and technology during the construction process. The topics covered include strategies for translating BIM to the field, integrating accountability, and how mobile technology is changing the game during the construction phase of a project. This chapter covers processes for quality control, installation validation, change management, equipment tracking, and inventory management. Lastly, this chapter covers how to create a real-time digital jobsite that is constantly connected with information being shared almost instantly.

Chapter 6: BIM and Construction Administration   BIM and construction administration is where information created and analyzed during preconstruction is put into use in the field. The combination of virtual environments with mobile-enabled site information has shortened the gap between information availability and response times. This chapter explores how to go from a BIM department to a BIM company. Additionally, this chapter looks at the various processes required of project teams in the field, document control, information clarification, sequencing, and project team training, and looks at the ways BIM and technology can reduce information processing times during the construction administration phase of a job. Lastly, it shows how to integrate best practices and capture knowledge sharing from one project to the next to improve the way an organization delivers a technology-enabled construction product.

Chapter 7: BIM and Close Out   Project closeout is often the last touch point with a construction consumer and is becoming increasingly important to deliver effectively. Many customers are becoming more informed on the value of as-built BIM and information for the life cycle of their project and are requesting new deliverables. While there may be projects that require a hardcopy set of as-built information and digital PDF sets, other customers have begun shifting to digital deliverables only. This chapter explores the artifact and constant deliverable strategy that better prepares a maintenance and operations team to update facility information.

This chapter also explores how to successfully deliver on promises made during the project planning stage and includes information on how to use technology to better perform project closeout, punch list issue resolution, and as-built capturing. Lastly, this chapter includes an overview of mobile applications and tools that make the job of closing out work easier and shows how to complete information migration requests into facility management or CMMS tools.

Chapter 8: The Future of BIM   Chapter 8 shares insights into what is in store for construction management. By looking at industry trends and new connected tools, enabled by new teams and collaborative processes, this chapter proposes an exciting and bright new future for the construction management industry. This chapter also shares information from other industries that have established knowledge management platforms with a focus on improvement and better quality, and it shows where many of these discoveries can be directly applied in the construction management space.

Addressing Change

So much has changed since the first version of the book that it only made sense to reinvent the focus of this version by taking into account the entire ecosystem of information management during construction. With information as the constant thread and enabling technologies such as BIM and mobile applications serving as the vehicle to provide a better way of collaborating around and distributing information, our goal for this book is to show the new "rounded out picture" of what BIM during construction management is being defined as.

While we know this book will cover specific technologies and tools, it is not intended to be exhaustive. By showing the bright spots as well as the challenges to using technology in construction management, we wish to add fuel to the fire of innovation that is happening within the construction industry. Just as BIM significantly impacted the industry, who is to say that there aren't innovative colleagues working together in a garage right now on the next application that will disrupt the existing toolsets and fundamentally change construction again? This is an exciting proposition, particularly for an industry that has not kept up with other major industry innovations over the last 40 years.

Lastly, we want to emphasize that change which creates successful outcomes requires better tools, different processes, and enabling behaviors. Construction management has indeed changed over the last five years, and it is our hope that it continues to change for the better over the years to come with a renewed focus on results and better information flow.

BIM and Construction Management

Why Is Technology So Important to Construction Management?

1

The construction industry is in the midst of a technology renaissance. BIM served as the initial catalyst for this period of innovation, but has now grown beyond "just BIM" to include innovations in many other areas such as mobility, laser scanning, and Big Data analytics among others. Supporting processes are changing as well. The construction industry is realizing that these new technologies don't fit into previous processes.

In this chapter:
The promise of BIM
The value of BIM in construction
Where the industry is headed

The Promise of BIM

Before the advent of BIM, the construction industry generally worked in silos, where each member of a project team looked out solely for his or her own best interests and the project took a backseat (or was in the trunk) to other priorities. Further compounding the isolation issue was the prevalence of the hard bid delivery method, which contractually and financially isolated team members from one another. Both the culture and this standoffish delivery method made for a litigious environment that was plagued with waste and cost overruns. According to the book *The Commercial Real Estate Revolution: Nine Transforming Keys to Lowering Costs, Cutting Waste, and Driving Change in a Broken Industry* (Wiley, 2009), by Rex Miller, Dean Strombom, Mark Iammarino, and Bill Black, the waste created by "simple efficiency and not-so-simple bad behavior" in the United States alone in 2007 was an estimated $500 billion. If we are to continue to function as a profession, we must ask ourselves, "Why should we ask construction consumers to pay for our mistakes?"

The promise of BIM is to build a structure *virtually* prior to physically constructing it. This allows project participants to design, analyze, sequence, and explore a project through a digital environment where it is far less expensive to make changes than in the field during construction, where changes are exponentially more costly. Today, this promise is becoming reality. An array of BIM software and mobile applications are delivering results that mitigate construction risk. Although some tools are more advanced than others, we are rarely at an impasse where some function is simply "impossible" and not able to be achieved through technology.

Where we find the majority of challenges nowadays in virtual building is that many teams fail to realize that the integration of team members creates significantly better outcomes. For example, subcontractors who are allowed to participate early in the scheduling process are able to leverage their expertise and share valuable information such as material lead times, crew sizes and installation methods that can create a more meaningful model simulation. Additionally, when a construction management team is allowed to participate in an architect's design review meeting, they are able to see what factors are important to the client and design team and use that knowledge going forward as they prepare to build. In this book, I acknowledge these best practices and propose a new way of evaluating technology and teams holistically by using *integrated teams* that are capable of keeping pace with the rapid introduction of available technologies to deliver better construction outcomes. As George Elvin states in *Integrated Practice in Architecture: Mastering Design-Build, Fast-Track, and Building Information Modeling* (Wiley, 2007): "Integration enables a team of designers and constructors to work together toward a common goal, allowing design and construction activities to unfold in the best way for the project, rather than locking them into separate phases required in over-the-wall delivery." It is this collaborative, project-focused approach that allows teams to function more efficiently and use BIM to get to better answers faster. Team integration moves the focus beyond individual needs and shifts it to how information-rich models can be used to explore options and scenarios that create better projects and remove risk.

BIM has evolved. The construction community is seeing a shift from the 3D or visualization aspect of BIM to workflow-specific tools that are being directly applied to solve real-world problems, such as installation verification, sequencing, and estimating. The industry dialogue is now moving to a general questioning of how we optimize the effective capture, analysis, and dissemination of information in real time to make projects more successful.

As a result of this shift in focus, existing tools are adapting and new ones are being created to address these challenges. The adoption of BIM into mainstream construction management practice has taken the typical constructs of what it meant to be a construction manager and transformed them into a new way of looking at how we work. We are now asking new questions such as:

- What else can we do with all this information?
- Who else can benefit from this data?
- How can we use models to enable better decision making?
- What is the right level of virtual augmentation on a project site to make our teams more productive?

It's an exciting time in the AEC industry because just as applications are improving, so are many of the technologies that support its use. Technologies such as cloud computing, which gives you the ability to use remote servers to process data from any web-connected device, and the accelerated growth of mobile and wearable hardware continue to shift the paradigm of practice in construction management for designers and builders alike.

Other changes are more incremental in nature. These improvements come in the form of better software features based on user feedback as well as enhanced stability of these tools, which increases productivity and reliability.

Finally, the constant stream of new ideas and improvements in the form of innovative tools and processes entering the marketplace continues to challenge the way in which teams work and build structures at a variety of levels. In the midst of all of this change is the promise of a better way of working collaboratively with more useful information to create value in the built environment.

Since BIM's introduction, BIM software has progressed with new features and applications. Likewise, BIM has forced many in the construction industry to evolve as well and challenge the way they previously thought about designing and building projects. As a result, the construction industry began investing in new and better technology. The rapid growth of new technology for the construction market is no coincidence. Construction hasn't kept pace with other industries in regard to automation and technological improvements over the last forty years, which has created fertile ground for new tools and products that offer better ways of working. Although innovation is encouraged, new tools require fast analysis and project testing before widespread adoption.

In the first version of *BIM and Construction Management*, I stated that BIM is not just software—rather, it is a process *and* software. Taking that one step further, we now see that successful BIM use requires **three** key factors:

- Processes
- Technologies
- Behaviors

These three components can make or break a project using BIM and technology. Think of these as the three-legged stool to the successful integration and use of BIM (Figure 1.1). Take one leg away and you are left with a pretty useless object that isn't good for much. So why are these three pieces so important?

Successful BIM Platform

Figure 1.1 Three-legged stool of BIM

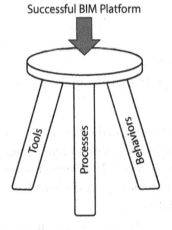

Processes

Construction management and many other engineering-focused firms tend to take new technologies and try to make them work in old processes. This approach creates waste by not taking into account the implications of the new tool and what existing processes and workflows should change that would make an outcome more efficient. A good example was the evolution of clash detection and resolution. As clash detection started to gain traction, many teams would host a number of meetings each week that involved the entire project team to coordinate among themselves using this new 3D environment. Although the technology was better, the process used was similar to what had been done before in a 2D coordination review. As a result, many users found the new process was not only inefficient but actually detrimental to a project's efficiency. Because team members were tied up in clash detection review meetings, response times for project-related issues increased. They were also burning through valuable time and found that their production declined steeply because of the lack of available hours. Nowadays, these meetings typically focus on two or three particular trades or scopes at a defined 2-to-3-hour timeframe to best use each team member's resources. Additionally, teams are now looking at ways of eliminating the clash detection process altogether by modeling in cloud-based tools that notify users in real-time when they create clashes.

These process shifts are critical to improvement, because they allow users to continually think of ways to improve and deliver work. In his book *The Spirit of Kaizen: Creating Lasting Excellence One Small Step at a Time* (McGraw-Hill, 2012), Robert Maurer states that "When you need to make a change, there are two basic strategies you can use: innovation and kaizen. Innovation calls for a radical, immediate rethink of the status quo. Kaizen, on the other hand, asks for . . . small, doable steps toward improvement." Successful BIM integrators realize that both large innovation and smaller process step changes are needed when using technology. Innovative change is driven by the speed at which technologies are deployed, and in order to stay relevant, you need to find ways to be nimble and look at these tools as fast as they come. Kaizen change calls for patient, iterative improvements to current tools and processes used and, at its core, require a cultural mind-set in order to work.

Keep in mind that, like a hammer or a saw, BIM is just a tool. Used with the right processes in place, BIM systems can create tremendous value for an organization. When new tools are combined with old processes, they can inhibit success as well as frustrate users. This is why it is so important to look at new tools as they become available for what they are and treat the investigation of the processes required to enable a new tool with the same rigor as that of the technology itself.

Technologies

The successful integration of BIM involves using BIM tools that work. Though this sounds simple enough, tools need to be explored further "post sales pitch." This means after the software or application salespeople have left the room, we need to ask, "Does this product improve our organization or way of working?" The strategy for how a team analyzes new technologies and selects them is important because it determines how nimble and responsive a team will be. The method for selecting tools in the construction industry typically falls into three approaches, each with different results.

The first strategy for selection and integration is the "pile on" method. In this approach, a company or organization looks at tools consistently as an *addition* to their current systems. The main hypothesis in this method is that the firm will begin by piloting the new tool and then look at how it interfaces with the company's other systems to see whether the product can meet its demands. If the tool looks like it is valuable and can be used, then the company begins a broader series of pilots that explore it further. The intention is that the new tool will "weave" its way into the fabric of the tools used within the company and ultimately the best tools will be used, while the others will fade away.

This method is the least painful of the three strategies, mainly because it is easy and requires the least rigor and thought. However, the constant addition of new tools creates confusion as to which tools are foundational and which are being tested. The pile on method rarely evaluates new tools against the current tools a firm is using. This type of diligence usually results in tools that overlap in functionality without a decision to remove one or another, until absolutely necessary. The pile-on method does allow for iterative or Kaizen-like changes to be made with little pain; however, a firm must be diligent about not selecting too many tools that inhibit the company's ability to perform.

The second strategy is a "swap out," or a direct replacement strategy. In this method, a company examines a new tool and its features and then looks internally to see which current tool or tools could be replaced. This one-to-one analysis allows for systems to be upgraded and consolidated. Direction on which tools are to be used and which aren't are usually clearer than with the pile-on method. This method also creates the ability to continually optimize the "toolbox" of a firm to stay relevant and competitive.

One of the shortcomings of the swap-out method is that the related processes and in-depth discovery of how a team works together takes a backseat to the feature comparison of each piece of software. Additionally, this method of selection is weaker against disruptive technologies that change the fundamentals in the way a company works, because behind the tools there is usually an established way of working. The improvement cycle in this methodology often follows industry trends, though this method does allow tool selection to be consolidated and the toolbox of an organization to be focused.

The third strategy is less well known but is now growing in popularity due to the rise of lean concepts and outcome-focused thinking. Using this method, known as the "process first" strategy, a team begins by looking at their current processes and then asks "How do we want to work?" This question requires "blue sky" thinking and assumes that the technologies needed to enable this new way of working will be there when they determine their optimal working conditions. This method of selection is more tedious and time-consuming than the two previous strategies and requires a significant investment of time and research to work. The outcomes from this effort vary, but many firms come away with a plan that includes input from a broad cross-section of their stakeholders. The difference is that the team understands the desired outcomes, and the selection of one tool versus another requires considerably less effort.

In this method, the litmus test of value is whether or not the tool aligns with the firm's vision. In some cases, no tools exist that support how a team wants to work. This situation is a risk of the process-first strategy; however, it is also fertile ground for customized solution development that meets the needs of the team. These custom solutions can be developed internally or with a third-party developer, or information can be provided to software vendors to develop and integrate into future releases of existing tools. This method of technology selection provides a framework for identifying tools that help a team reach its desired end state, because it allows the most flexibility in a rapidly changing environment and limits the "analysis paralysis" stage that many organizations face when analyzing tools from too many perspectives.

Unless a firm truly hasn't changed tools in some time, it will typically use one of these three methods or some combination. Whether the methodology of selection was purposeful or less rigid, a firm that wants to continue to adapt and improve should look at the way it analyzes and selects tools. Doing so determines the speed and efficacy of that company to stay at the forefront of technology and market trends.

Overall, BIM in construction is seeing a trend of consolidation in quantity and a focus on cross-platform integration. Some vendors are rising to the call of interoperability, application programming interfaces (APIs), and open source information sharing that limits redundancy and starts to create interesting new ways of using BIM information. This continued improvement in BIM software can largely be attributed to user communities and feedback. Whether that feedback comes from online forums, consumer councils, or involvement in industry organizations and committees, the lifeblood of improvement in BIM relies on users in our industry to take an engaged stance in the future iterations of existing tools in these venues. Just as important is the willingness to be "sold to" by new companies with new ideas to support a dialogue and cultivate a culture of innovation and advancement within the construction community.

Behaviors

Of the three key components to successfully integrating BIM, behaviors are the most difficult to change. As Scott Simpson of design firm Kling Stubbins says, "BIM is 10 percent technology and 90 percent sociology." The core of BIM is far more than updating software—it is a cultural shift in the mind-set in the way construction management teams collaborate. So, what do we mean when we say "behaviors"? When we consider what makes BIM work within a construction project, the core component becomes *enabling behaviors*. Think about it. Would you rather work with a team that is excited to work in a cutting-edge environment—or a team that is overly skeptical and limits further progress by being closed-minded?

Not a tough decision to make.

Teams need to fully realize that a future forward mind-set is just as important as the technologies and processes behind it. Those who misunderstand this principle will quickly find themselves irrelevant in the design and construction market. As the philosopher Eric Hoffer says, "In times of change, learners inherit the Earth, while the learned find themselves beautifully equipped to deal with a world that no longer exists."

Although we have discussed the importance of personal behaviors, it is also important to note that organizational behaviors can impact the successful integration of technology as well. A company that has a culture of innovation and a nimble attitude to begin with will create a persisting dynamic where change is a constant and improvement and analysis are to be expected. Conversely, an environment that is resistant to change and that stifles innovation will become exponentially more difficult to create that enabling dynamic that supports the successful analysis selection and use of the right tools that may translate to process changes.

Behaviors Matter

Construction management firms are facing an increasingly competitive environment all over the world. This is particularly true for large projects, where significant effort is required and large amounts of revenue can be made or lost. In many of these projects, joint ventures (JVs) are used to take the best of what both teams have to offer as well as spread out the risk, bonding, and insurance requirements. It is important to note that when JVs are being created, individual teams are selected based on various factors, including their experience, portfolio, client relationship, technological capability, availability, and behaviors. Why behaviors? Well, these projects often carry a significant amount of risk, not only as it pertains to the construction project but also as two or more companies with different cultures begin working together. For this reason, teams with the right enabling behaviors often find themselves as a desired partner, whereas teams resistant to change often find themselves left on the sidelines.

One of the main themes in Finith Jernigan's *Big BIM, little bim* (4Site Press, 2008) is the concept that truly successful BIM is much more than just BIM software (little bim); rather, it is the assemblage of the tools, processes, and behaviors (BIG BIM) required to make BIM truly effective. Just as BIM tools are becoming more collaborative, so must our behaviors and mind-sets. We as an industry have a significant opportunity to capitalize on what has the potential to revolutionize the way construction is delivered going forward by shifting our attitudes and mind-sets to more enabling behaviors.

The Value of BIM in Construction

The value of BIM in construction comes in many shapes and sizes. Whether it's the ability to save time through automated functions, eliminate the need to travel to a meeting, or save money because better information is available earlier to make cost-effective decisions, they all have the same focus: results.

It's hard to imagine an area of our daily lives in which technology doesn't affect us, particularly in the workplace. The same is true within the construction industry. The advent of BIM and the rise of application-based technologies have opened doors and arguably created one of the most exciting new dynamics since Microsoft Excel. Over the last 50 years, the construction industry has had just a handful of notable technological innovations compared to other industries. Granted, there were many innovations in material research, installation methodologies, and energy efficiency, such as prefabrication, eco-friendly materials, and green building design. However, the technologies used by project teams for construction management remained largely the same. Now, innovation is becoming a part of the way contractors deliver their work and differentiate themselves from their competitors. As a result, we are starting to see a healthy ecosystem of supply and demand for ever better tools between technology vendors and construction management firms willing to invest to drive efficiencies, as is evident in the rise of contractors adopting BIM technologies (Figure 1.2).

Contractors' Current and Future Expected BIM Implementation Levels

Source: McGraw-Hill Construction 2013

■ Users at Low
   BIM Implementation
   (Less Than 15% of Projects)

▨ Users at Moderate
   BIM Implementation
   (15%–30% of Projects)

■ Users at High
   BIM Implementation
   (31%–60% of Projects)

▨ Users at Very
   High BIM Implementation
   (More Than 60% of Projects)

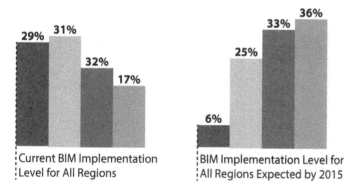

Current BIM Implementation Level for All Regions: 29%, 31%, 32%, 17%

BIM Implementation Level for All Regions Expected by 2015: 6%, 25%, 33%, 36%

Figure 1.2 Expected growth trends of BIM

BIM in construction management has a unique history. It is important to understand this unique evolution in order to best understand its value and trajectory (Figure 1.3).

BIM as we have come to know it is largely based on object-based parametric modeling technologies that were developed by the Parametric Technologies Corporation in the 1980s (source: *BIM Handbook*, p. 29). BIM for the construction industry was commercially available as a tool in the early 1990s with the ability for computers to handle the size and processing requirements of 3D CAD models. The acquisition of Navisworks (Formerly: JetStream) by Autodesk in 2007 served as a catalyst for BIM adoption among contractors because of its ability to integrate multiple BIM filetypes. As BIM became more mainstream from 2007–2010, the series of follow-on applications, services, and hardware that were associated with this exciting new tool grew significantly. This surge in quantity of BIM-related plugins, add-ons, and applications created the two new dynamics that we outlined previously. The first dynamic was the early stages of the technology "renaissance" that moved the topic spotlight to the construction space and where it stood in its use of technology. The second dynamic created was the challenge imposed upon construction firms to select the right BIM tools that worked together to create value. This phase of BIM history was largely viewed as the beginning stages of BIM, and it quickly brought about a call to action between these tools on the topic of interoperability and the free exchange of data between systems, which continues to be a discussion point today.

The percentage of companies using BIM jumped from 28% in 2007, to 49% in 2009, and to 71% in 2012.
For the first time ever, more contractors are using BIM than architects.

Source: *The Business Value of BIM in North America: Multi-Year Trend Analysis and User Ratings SmartMarket Report*, McGraw-Hill Construction, 2012.

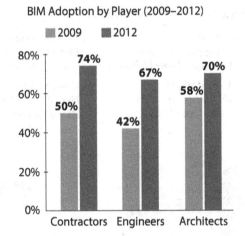

Figure 1.3 Growth of BIM adoption since 2007

Some early adopters began integrating BIM technologies within their firms with the intention of differentiating themselves from their peers. Many companies tried to label their unique brand of BIM as something special that they alone understood and that no one else had. Whether it was key staff, project experience, or custom-developed tools, the need to stand out from the growing BIM crowd became increasingly important. The adoption of the tools often took the "pile on" approach as described earlier in this chapter, and it seemed that competition was driven solely on the basis of who had the latest new tool. This drove many early adopters to question the context and value of BIM tools and workflows in order to make them more competitive. Ultimately, this served as a catalyst for broader thought and deeper questioning.

Another aspect contributing to the rise in quantity of innovators and early adopters can be attributed to BIM being requested by forward-thinking construction consumers such as the General Services Administration (GSA), United States Army Corps of Engineers (USACE), Disney, Google, Coca-Cola, and other large-construction consumers. In order for construction firms to stay relevant to these customers, the need for a compelling technology approach rooted in proven deliverables and consistent results became more important.

Many technologists predicted that it would be decades before BIM took hold as a tool and process. Although initial efforts were slow and followed the trend line of a traditional technology integration cycle (Figure 1.4), BIM has since taken the AEC industry by storm and firms that are now using BIM have surged from 28 percent use in 2007 to 74 percent in 2012 (source: *BIM SmartMarket Report*, "Business Value of BIM in North America") and has quickly shifted from early adoption stages to the middle and late stages. As the dust settled from the initial excitement and optimism, early adopters were replaced in number by the early majority, who investigated these new tools at a much deeper level than surface promises. Ultimately, these industry experts and analysts began to weigh in on which tools provided clear value and delivered on their promises and which ones did not. Many communities and organizations were created as a means to capture this information.

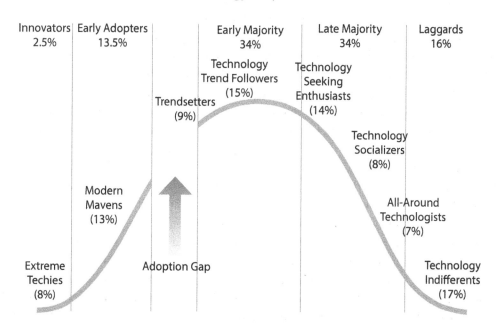

Source: *The Nielsen Company and Cable & Telecommunications Association for Marketing (CTAM)*

Figure 1.4 Traditional technology adoption cycle

From 2007 to 2012, the factors influencing the use of BIM shifted. In 2007, the second-largest factor in increasing use of BIM was owners demanding it on their projects (Figure 1.5). However, in 2012 owners' demand for BIM fell to the number four slot as interoperability, functionality, and clearly defined BIM deliverables took its place (Figure 1.6). Although owner requirements are still a major consideration in BIM use, we now appear to be at the crest of BIM adoption in the construction industry. Proof of deeper questioning and analytical thinking that is consistent with trends of the more careful early and late majority adoption cycle group continues to surface as more professionals from all over the world explore the depths and potential for BIM in construction.

As with the traditional technology adoption cycle, the innovators, early adopters, and trendsetters pass the baton to the early and late majority, which is largely where BIM is in its technology adoption life cycle currently. The early and late majority users, characteristically, are more analytical than "techies" and have had a significant impact on BIM use, with more to say about workflows and the quality and organization of data than their predecessors. There continues to be a big focus on interoperability between tools and a more in-depth look at the value case of large amounts of data that host or link to parametric (model) elements. However, the industry is still defining the value of BIM information as a whole. This presents many opportunities for future innovations that will go through additional technology cycles. Many industry organizations host presentations and discussions that have moved beyond introducing BIM as a concept to more detailed analysis on planning, organizational structure, and process change dialogue.

This is a very exciting time for BIM and technology within construction. We are now seeing new tools and processes come to market that are focused on value that align to processes and consolidate functions. Examples are the new version of Vico Office, which integrates time and cost functions; Autodesk 360 Glue, which eliminates the need for working on files in a LAN or WAN configuration; and Bentley's ProjectWise Construction Work Package Server, which integrates the model, schedule, and task planning.

So where is the value in BIM now?

Simply put, the value in BIM is still in the *information*. However, the value is more known than before and it is now this better connected, simpler more results focused approach that is guiding the use of this information. When we think about the potential uses for a virtual model that is able to contain information about each individual door, roof top unit, slab, and window, we begin to understand that the implications for estimating, scheduling, trade coordination, and installation are still profound. The design and construction industry will continue to become more efficient at creating and using models. As a result, teams are exploring ways of using the data and information produced from models to eliminate waste in redundant data entry and points of input as well as to spot trends, patterns, and issues that we simply were not capable of capturing before the introduction of BIM.

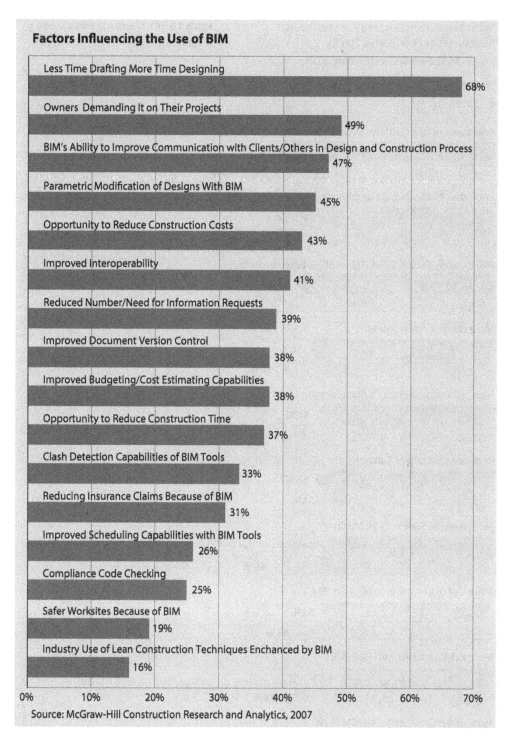

Factors Influencing the Use of BIM

Factor	Percentage
Less Time Drafting More Time Designing	68%
Owners Demanding It on Their Projects	49%
BIM's Ability to Improve Communication with Clients/Others in Design and Construction Process	47%
Parametric Modification of Designs With BIM	45%
Opportunity to Reduce Construction Costs	43%
Improved Interoperability	41%
Reduced Number/Need for Information Requests	39%
Improved Document Version Control	38%
Improved Budgeting/Cost Estimating Capabilities	38%
Opportunity to Reduce Construction Time	37%
Clash Detection Capabilities of BIM Tools	33%
Reducing Insurance Claims Because of BIM	31%
Improved Scheduling Capabilities with BIM Tools	26%
Compliance Code Checking	25%
Safer Worksites Because of BIM	19%
Industry Use of Lean Construction Techniques Enchanced by BIM	16%

Source: McGraw-Hill Construction Research and Analytics, 2007

Figure 1.5 Factors influencing the use of BIM: 2007

Most Important Factors for Increasing BIM Benefits (2009 and 2012)

Source: McGraw-Hill Construction, 2012

Figure 1.6 Most important factors for increasing BIM benefits: 2012

■ 2012
░ 2009

Improved Interoperability between Software Applications

68%
79%

Improved BIM Software Functionality

64%
78%

More Clearly-Defined BIM Deliverables Between Parties

62%
70%

More Owners Asking for BIM

58%
67%

More 3D Building Product Manufacturer Content

56%
65%

Reduced Cost of BIM Software

56%
54%

More Internal Staff with BIM Skills

54%
69%

More Use of Contracts to Support BIM

54%
62%

More External Firms with BIM Skills

48%
66%

More Entry-Level Staff with BIM Skills

46%
54%

It's Not Just 3D

It is important to understand that though most have come to understand BIM as a 3D tool, it is also an information-rich database that links to and controls those model components; this is often referred to as "parametric modeling." There is significant value in the three-dimensional aspect of BIM, but its ultimate value lies in the ability to aggregate, edit, sort, and compile this information to drive at better answers to design and construction questions such as "What is the best sequence to install this piece of equipment?", "How much square feet of raised flooring do we have in this facility?", and "What are the parts I need to build this manufacturing plant addition?" If a 3D element is in the model, chances are a whole host of information is behind it that can be used in a variety of ways.

Where Does BIM Play a Role in Construction Management?

BIM continues to redefine the way the construction sector builds and works together. The core value of BIM that the construction industry should be aware of is the ability to take model information and extend its use by giving it meaning for other related workflows and processes. These workflows include impacts to basic functionality such as estimating, scheduling, logistics, and safety. These new capabilities have opened doors for faster population of data into these systems to deliver work earlier, safer, and with better quality.

Whereas data input is becoming more efficient, there is also an industrywide push for constant interconnectivity of data within these systems to behave more like an ecosystem of information in lieu of one-off file exports or imports. In other words, as Daniel J. Boorstin says in his book, *The Republic of Technology: Reflections on Our Future Community* (Harper & Row, 1978): "Technology is . . . fun, but we can drown in our technology. The fog of [too much] information can drive out knowledge." In essence, too much information can create risk by diluting the focus of a project. Information is critical in project management; however, when it becomes too cumbersome to manage or decreases effectiveness through overanalysis, it can be detrimental.

Lastly, we are seeing a big shift in collaboration methodologies and an increase in the frequency of information sharing. New tools such as box.com, Dropbox, Egnyte, Newforma, and other web-based file-sharing platforms are making the process of information sharing easier. Additionally, the cloud has significant potential to impact the way construction is delivered and managed, from rigid and limited to flexible and scalable. In the construction market, BIM and other related technologies have a significant place at the table for almost every phase of construction, and adoption continues to increase at a rapid pace (Figure 1.7).

Levels of BIM Adoption in North America

Source: McGraw-Hill Construction, 2012

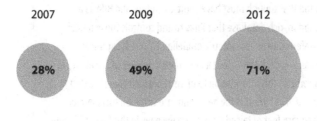

Figure 1.7 Increase in BIM adoption 2007–2012

Team Engagement

When it comes to communication and planning, the use of technology is critical in passing and sharing information. The use of BIM on construction projects is a great enabler for better team engagement. To ensure the successful participation of a team, BIM-enabled projects require that project plans be created to establish expectations and iron out project details upfront. The biggest inhibitors to engagement are confusion, complexity, and lack of communication. For this reason, it is paramount for each team member to understand the details of a project.

To best understand the use of BIM, team leaders need to create plans that describe the tools that will be used, acceptable file formats, and when team members can expect to receive information, among other factors (Figure 1.8). These plans will be covered in Chapter 2, "Project Planning," but note that the trick to solid team engagement is making it last throughout the project. It's easy for the excitement of a new project and a new set of challenges to make a team come together. However, successful team leaders know that keeping everyone engaged as active participants is key to project success.

Project Pursuit and Business Development

Owners continue to become increasingly informed about BIM and technology in construction, and many owners want to take more active roles in understanding what the teams proposing on their projects are capable of and where they have experience or have driven innovation. Particularly of interest to owners is the ability to leverage models throughout the design and construction process to remove unknowns and mitigate as much risk as possible in their projects. This can be accomplished in a number of ways, such as enhanced visualization (Figure 1.9) to understand material choices more accurate estimating to eliminate cost overruns, more detailed sequence and scheduling analysis to reduce negative schedule impacts, and creating safer job sites by laying out more easily readable 3D site plans. Lastly, many owners are beginning

16

CHAPTER 1: WHY IS TECHNOLOGY SO IMPORTANT TO CONSTRUCTION MANAGEMENT?

to use as-constructed models for life-cycle efforts to better inform operations and maintenance processes and carry information about the final project into asset and facility management.

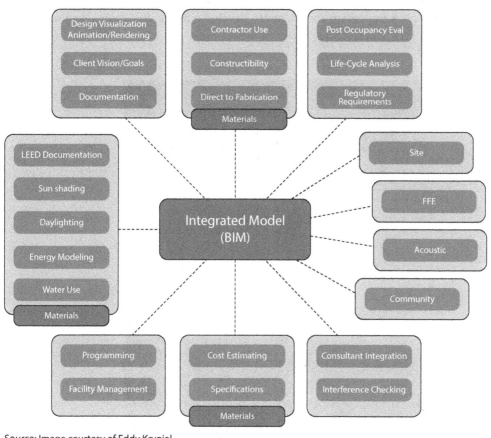

Source: Image courtesy of Eddy Krygiel

Figure 1.8 BIM has multiple stakeholders, so defining team members' responsibilities is critical.

In the initial project pursuit phase, it is important for construction management teams to articulate their tools and processes as well as their desired outcomes and where applying a certain technology will either remove risk, create value, enhance building performance, or better enable collaboration and communication. This can be accomplished by articulating the expectations and intentions of the team in BIM execution plans and information exchange plans before the start of the project.

Teams should demonstrate where they have analyzed an owner's needs and selected a tool that has created value. In applying BIM and technology in construction management, it is important to remember that not *all* tools are meant for *every* project. For example, the suite of tools selected to build a new high school will be different from those BIM tools selected to build a new cancer research center. The information required to build each is just as different as the facilities themselves. Though some of the information will be consistent between each project, it is important for a team to develop a methodology of selecting the right tools.

PHOTO COURTESY: LEIDOS ENGINEERING

Figure 1.9 Project visualization example

As teams develop expertise in applying BIM in a variety of projects, the tools and processes become clearer and teams recognize opportunities to innovate as a result of enabling behaviors and cultures. For instance, a team that has become proficient at model-based scheduling may find a project owner who requires the integration of cost into the project. Using their foundational understanding of model-based scheduling, teams may be able to integrate costs into that schedule as well. Innovation in the BIM process is critical to advancing both the industry and individual firm performance.

Planning for BIM Success

A myth in BIM is that a team should start immediately using BIM during the preconstruction phase. Jumping into BIM without having an execution plan or information exchange plan severely limits a team's ability to perform. During construction is not the time to work out how, if, and when systems should share and link to data throughout your project. In our experience, the success of BIM is determined by how well BIM use is planned for and communicated among a project team well before any preconstruction or construction activities begin.

The analytical nature of the tool and the powerful information that is tied to building information models creates a tremendous opportunity to capitalize on the use of BIM throughout the building process. These uses include estimating, quantity validation, sequencing simulation, design updates, version-based model analysis and owner/user group visualization, prefabrication detailing, material impacts, life-cycle energy usage, and life-cycle cost estimates that include utilities and operations, and so on. However, it is critical for a team to determine up front which uses and responsibilities make for a smooth project.

Using Contracts in Planning

Just as BIM and technology should be planned for, a best practice in using BIM is to contractually establish expectations, files to be used, frequency of distribution, and quality control checks. An array of BIM contract addenda is available from the American Institute of Architects, Design-Build Institute of America (DBIA), and Associated General Contractors of America, and more are detailed later in this book.

Standard BIM planning tools are available in the industry as well. These standard plans include the Penn State BIM Project Execution Planning Guide, the DBIA BIM Manual of Practice and Checklist, the USACE BIM Project Execution Plan, and the GSA BIM Guide, which are all available at no cost. Additional resources are available for purchase. These plans and guides are meant to act as a template for teams to detail how they will work to deliver a BIM project. In the case of the USACE and the GSA, their plans are an owner-driven standard project requirement that each team must complete and get approved. This creates consistency in deliverables and language while ensuring that the basic owner's requirements are met.

Some project teams decide to forgo these boilerplate templates and develop their own BIM execution plan or information exchange plan and include them as attachments to subcontractor agreements. Additionally, some customize their own plans to more specifically address the needs of their project, team, and contract formation and the desired end state. These customized plans are typically built around a number of best practices and years of experience. I recommend that early BIM adopters begin with the boilerplate documents to familiarize themselves with their parameters and refine these documents before creating one from the ground up.

Another tool that is growing in popularity is the memorandum of understanding (MOU), which is typically found in project types like Design-Build, IPD, and other forms of integrated delivery. The MOU is a charter document that maps out the intentions of the project team, the high-level deliverables of a project, resolution methods, and goals. A project team signs this agreement and uses it as a tool—and quite often as a contract addendum that needs to be adhered to. MOUs are typically used if little project information is available at the time the project is being pursued.

Regardless of the plan used, you must consider a number of pros and cons when structuring a team. BIM aligns to the construction industry at large that is headed to a collaborative and more integrated team approach, where the project is the focus and team members determine their value and use tools that enable them to make better decisions together.

Scheduling

Schedules in construction are meant to clearly define how projects will be assembled with defined activities and sequencing logic that establishes durations and directs the overall flow of progress on site. Schedules have traditionally been created in a vacuum by one or more dedicated scheduling professionals, whose role on a project is to take an educated guess based on past performance and other available industry data to generate a project timeline. Although this method of scheduling has long been considered the golden standard, it is by no means effective. According to *The Commercial Real Estate Revolution: Nine Transforming Keys to Lowering Costs, Cutting Waste, and Driving Change in a Broken Industry* (Wiley, 2009), by Rex Miller, Dean Strombom, Mark Iammarino, and Bill Black,, traditional project management schedules are wrong 70 percent of the time. So why do we keep using them?

One answer is that there simply is not another available solution. Yet, many contractors are now realizing the value of the integration of the model with schedule information. From this effort, firms are producing collaborative simulations with subcontractors to validate and ensure schedule accuracy as well as create efficiencies. The integration of schedules into BIM has been coined "4D" as well as "model simulation" studies and "sequencing animations." Model information for scheduling can be used a number of ways. Because virtual elements or geometric components are visible in a model environment such as Navisworks and Synchro, they can be animated. By linking these assemblies to schedule data, a video of the project's construction can be simulated. Additionally, these linked simulations can also detect incorrect schedule logic through sequenced clash detection. These simulations visually highlight issues such as equipment being set on raised pads prior to curing being completed, or beams appearing in thin air without supporting columns or superstructure.

BIM in scheduling continues to create a higher level of project clarity and has proven to be an effective means of communicating with a team visually for how a project is going to be put together. The tools are growing in sophistication, and many are starting to provide features such as the ability to slice models into phases and schedule tasks as they would be constructed for Location Based System (LBS) or Advanced Work Packaging System (AWPS) work, as well as introducing optimization features such as line-of-balance schedule views (Figure 1.10).

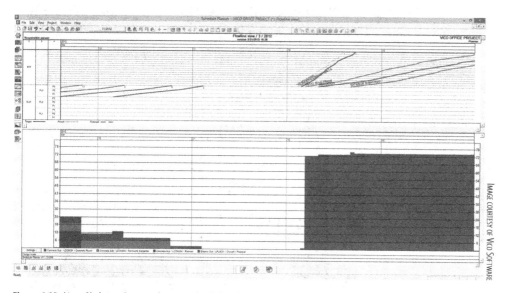

Figure 1.10 Line of balance, Gantt, and resource-loaded schedule view

BIM and scheduling have taken on a more in-depth relationship as many tools now directly link to each other. These tools, including Navisworks, Synchro, and Vico, make the process of updating and editing both the model files and the schedules much easier. Additionally, schedules can be loaded with cost and crew information, which can be validated before work starts to ensure the correct amount of materials and crew size. Later, in Chapter 5, "BIM and Construction," will go through the process of integrating schedules into BIM files and what tools are available in the market. As tools become more collaborative and the industry continues to shift toward more lean scheduling methods, traditional scheduling methods will be seen less and less.

Logistics

Traditionally, site plans have been used as the vehicle to coordinate site movement, safety, laydown, storage, equipment use, and access efficiency. BIM combined with geographic information systems (GIS) technology has changed the way we create site plans by adding a level of detail, particularly in more vertical construction. The integration of intelligent equipment and objects that communicate with humans will make our jobsites safer and more productive. Beyond site logistics, the integration of wearable technology—such as Google Glass, Apple Watch, and the Oculus Rift virtual reality goggles—with BIM and virtually hosted information is becoming a rising trend, with massive potential for spatial validation, onsite access to information, safety, and augmented reality (Figure 1.11).

Figure 1.11 Apple Watch

Site safety risks are an area in which technology will continue to take a role in mitigating. BIM and technology will continue to be looked on as a resource for the mitigation of risk and better visualization of onsite conditions. Actual site conditions layered with BIM data and real time equipment locations could make for safer jobsites around the world. Lastly, the acquisition of information earlier in a project life cycle allows team members to make issues known and addressable at an earlier stage in a project and to design in safety to the way a project is constructed.

Estimating Cost

BIM-derived estimating (also known as 5D) has been long believed to be the "golden goose" of BIM in the preconstruction phase of a project. Conceptually, BIM-derived estimating uses the database behind a building information model to either directly link those model components to unit cost or to cost assembly recipes to produce an estimate. For example, a model that contains 10,000 sq. ft. of drywall would link to a unique cost recipe including such information as the crew size, hourly rates, material costs, and productivity rates. This enables the model to define how long it will take to install the amount of materials being pulled from the model. In essence, the schedule becomes a product of how much "stuff" you are going to build in a particular order (Figure 1.12). This is unique compared to previous methods, where the schedule and the estimate often operated as two autonomous files and rarely were tied together. Products such as Trimble's Vico software allow for the building information model to become more integrated as it removes the process of linking disparate schedules to the model and in effect contains both the building components (3D), schedule (4D), and cost (5D) information within one tool.

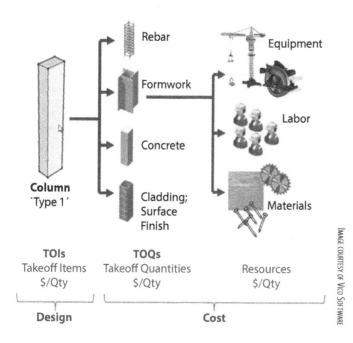

Figure 1.12 5D data flow

BIM estimating has grown significantly in sophistication and ease of use since early iterations of the software. Now tools have begun to make the process of BIM estimating more seamless, allowing for the integration of "non-model"-related data such as 2D PDFs and CAD drawings to be used in the takeoff process. Additionally, aspects of scope, such as general condition costs and equipment leasing, are now able to be integrated into an estimate, whereas before this they would have to be done in two separate environments. The industry now has a more robust understanding of how to accomplish model-based estimating, and many best practices are available from vendors, user groups, and industry associations to help tackle estimating modeled versus placeholder or assumptions. Model-based estimating still requires a significant investment to set up the cost databases and processes used to estimate and update BIM-derived information. However, many firms are seeing efficiencies in updating and accuracy.

Other innovations such as cloud-based model collaboration is starting to have a unique impact on a team's ability to complete estimates. Tools such as Onuma System (Figure 1.13) allow multiple users to collaborate in real time in the cloud with the same datasets. Collaboration in construction has started to shift beyond the model. It is web-based and real time, and it involves input from several team members and aims to speed up work. Enabled by technology, the process of collaborative estimating and scheduling is beginning to take hold.

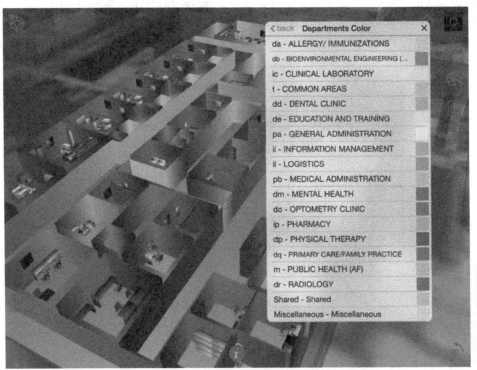

Figure 1.13 The Onuma System on the iPad

Constructability

Constructability is a project management technique that reviews construction logic from beginning to end during the preconstruction phase to identify roadblocks, constraints, and potential issues. The use of BIM during constructability for spatial coordination was a catalyst for the rapid adoption of BIM in the construction industry. In an interesting progression, the value of using models to coordinate system and structure layout has been transformational in how the construction industry now handles conflict resolution between systems. Previously, spatial coordination issues were addressed over light tables by individual reviewers. Now, models are reviewed in 3D, collaborative environments to best understand the issues and then resolve them (Figure 1.14). Often the resolution of conflicts can be achieved in the same environment, significantly reducing the amount of phone calls, e-mails, and other coordination efforts. This area has garnered perhaps the most sophistication in the construction management space as many contractors now use BIM as a best practice in coordination of their projects, regardless of project value, type, or size.

Figure 1.14 Model coordination review meeting

The process refinement of constructability, also known as conflict detection and resolution, has grown in scope and sophistication as well. It is not uncommon to see clash detection reviews now include installation and maintenance clearance objects, or "blobs," in front of equipment. These blobs allow team members to understand that even though there is nothing physically in the space, the space is actually needed for accessibility and maintenance purposes (Figure 1.15). For example, a mechanical subcontractor may place a 12" clearance blob in front of a piece of equipment to ensure that the filters in a VAV box can be replaced. Overall, constructability and systems coordination in BIM is a known and accepted process in preconstruction coordination, with many firms using BIM as a best practice.

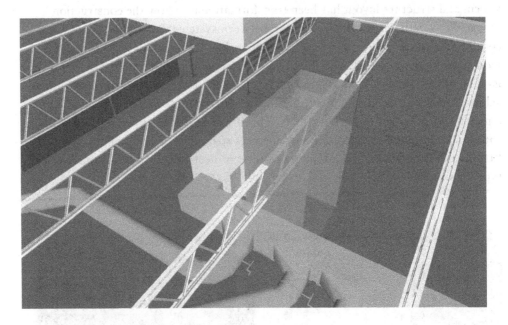

Figure 1.15 Clearance object in front of equipment

When you think about it, clash detection ultimately represents a *failure* in the design process. In the past, BIM teams were limited by technology. One of the main limitations was model coordination lag due to the inability to transfer large model files over the Internet. Although model file sizes vary, many projects can range from a couple of megabytes (MB) to some as large as several gigabytes (GB). For this reason, the ability to work on local files over the tried-and-true local area network (LAN) ceased to become an effective solution if the team was working in different places. Unfortunately, the workflow became a "snapshot" approach to the BIM process. Because information could not be shared with team members fast enough, design teams worked on an instance of the main or central model in a vacuum on their local devices or networks and would then reconcile their updated model with a central or composite model at a later time only to learn that there were now a whole new number of issues to resolve.

There was a major roadblock to this process. As multiple team members all worked in their own instances of a model, conflict resolution became a time-consuming process and multiple "speed bumps" in the road caused pause for coordination once models came back together. Many teams found resourceful solutions. On such solution was co-location; with this approach, the entire project team would physically sit in the same room. Another solution was staggered design development phases, which would let a specific discipline "own" the model, add information, refine their design, and then pass it to the next user. These methods were effective to some extent, but for many projects they were unsustainable in the long term. Co-location, for instance, typically had costly travel expenses. As for staggered phases, project timelines often didn't allow the luxury of waiting for design development to be passed from one set of users to another.

As a result, software vendors and application providers have begun to consider other solutions. One is to co-locate virtually by creating an environment where the project team, regardless of discipline, can "plug and play" via the cloud. This solution is promising and allows model file transfers and real-time coordination, and enables geographically disparate teams to work together. Over the last five years, there has been a narrowed focus in collaboration in the cloud. To some, cloud computing means a remote hosting offering; for others, it involves using a network of remote servers to accomplish such tasks as data crunching or *Big Data compilation* (where large amounts of data are sorted, filtered, and analyzed to identify patterns and trends). As it pertains to BIM, the use of cloud-based model collaboration through a virtual desktop environment (VDE) has opened the doors wide for an exciting new collaborative platform as well as a tool to complete analysis and design faster.

This new network structure creates a platform of interaction that often extends well beyond the project team into other areas, such as the ability to use crowdsourcing and get third-party insight into construction issues. Because of this new platform, we are starting to see real-time clash detection occur as models are being developed. This new ability frees up time from the traditional constructability review and resolution process and puts users on a course for better effectiveness and time management. Tools such as Autodesk BIM 360 Glue (Figure 1.16) introduce a real-time clash alert that lets users know when the systems being modeled are creating conflicts with other systems.

Analyzing Data in BIM

The ability to analyze the large amount of data available in BIM or that is gathered through the design and construction process creates a broad array of possibilities. Many systems have now been developed with a particular focus on aggregating this data and making it more meaningful during design and construction. Some of these tools focus on work such as equipment management, safety, inventory tracking, issue management, document accessibility reports, and an array of other solutions that aim to simplify, automate, and streamline work that happens in the field.

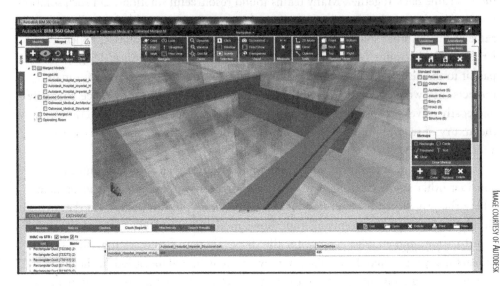

Figure 1.16 Autodesk BIM 360 Glue

The topic of analysis is almost too broad, and for this reason, we will focus on tools that are used during construction that achieve meaningful metrics that can assist a construction management team in becoming more productive, safer, and more connected. The field of analytics is relatively new in construction. Tools such as Sefaira (a carbon footprint calculator), STAAD (used for structural analysis), Trane TRACE (a computational fluid dynamics tool), and many other data-intensive applications have largely focused on the design space. The value proposition in design is much clearer in part due to the availability of model data that can connect to current software in-house for analysis functions with a clear end product—that is, structural analysis, energy use, carbon footprint, daylighting, and so forth.

In construction, we are just now beginning to see the integration of analytic tools into the way we work and coordinate and in the way we build. There is an exciting value proposition in some of the data and what it will tell. Particularly of interest will be the constructs Big Data will challenge and how teams validate reaching their goals through measured outcomes and navigate quickly through volumes of information.

Construction projects continue to become more complex. The requirements asked of construction managers are growing in the areas of safety, reporting, and information management. Adding to this complexity is the rise in quantity of available information from relatively new technologies such as BIM. Although this presents unique opportunities, it also creates new challenges to managing data. The management of this data is directly related to a certain amount of risk. Managing information like safety reports, material inventory, subcontractor performance, schedule updates, accounting, and quality control have traditionally been in the scope of construction management firms. Now as the construction industry has largely accepted the value of BIM, it is now faced with other issues. Who is responsible for managing the information in the model? What other value can be extracted from its use? The future landscape of information analytics in BIM is a promising field, but there is a large amount of work that can be done in this space to develop exciting applications that help teams and our industry make better decisions.

Designing for Prefabrication

The promise of prefabrication is nothing new to the construction industry. Beginning as far back as 1624 with the use of a wooden panelized building to house fishing fleets to more of "prefabrication" as we have come to know it with the Aladdin Readi-Cut homes in 1904, the art of prefabricating offsite has continued to be of interest to builders as a means to build faster and more efficiently (source: `http://oshcore.com/thesisbook/Chapter%201.pdf`). There are many upsides to prefabrication, or "prefab":

Advantages of a Controlled Environment   Work is completed in a controlled environment, where tools, workspaces, and sequence are predetermined, thus making for a safer project site as the use of ladders, lifts, and other on-site equipment is limited and work is performed in assembly line style.

Avoiding Delays Caused by Bad Weather and Other Unforeseen Conditions   Production environments are often indoors or covered, so weather or site conditions don't inhibit the construction process and work can continue to progress in inclement conditions.

Efficiency of Assembling the Parts On Site   Although there is debate about the speed of initially constructing the "pieces," there is no doubt that the speed at which the "parts" can be assembled on site is far faster than site installation. There is a significant time savings in installing parts such as walls, floors, ceilings, and roof elements that come in modular components.

Prior to widespread adoption of BIM, these prefabricated components were coordinated and "sliced" using CAD, which had limitations in fully visualizing how 3D structures were to be built. The advent of BIM has made prefabrication much more dynamic and in general follows this process:

1.   The overall building design is completed by a design team and then sent to a fabricator for review.

2.   Components of the building are then "sliced" into buildable and, more importantly, transportable chunks.

3.   Next they are sent to engineering for material ordering and equipment procurement and sign-off.

4. The building "slices" are then laid out onto assembly lines where the various components are arranged in the most approachable fashion to construct. Often the construction of the pieces follows the project construction schedule.

5. After components are constructed, they are packaged up, numbered, and shipped to the site.

6. Components are then installed on site (Figure 1.17).

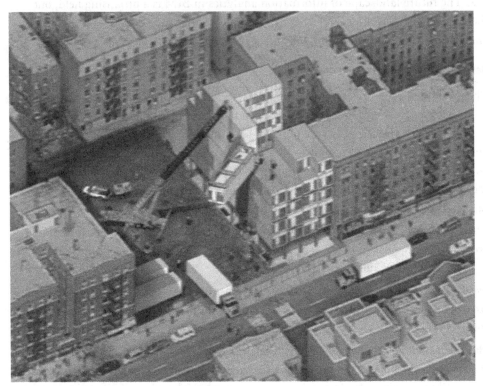

Figure 1.17 Prefabricated project, "The Stack" in New York City

The introduction of BIM into this process allows for design teams to actively participate in the review and design of the modules to be prefabricated. By showing where these slices should best occur, design teams are able to optimize modular units with the fabricator for the layout and sequence of how the buildings' parts are built and installed.

Future industry speculation in this space alludes to a "print building" button that would eventually allow design and construction teams to send sliced model files direct to the fabricator, where building components can be constructed in near real time. The introduction of large scale 3D printing is a disruptive technology that could significantly impact the way buildings are built. Loughborough University as well as some private companies such as WinSun have done some fascinating work on this subject.

The ability to "print" building materials such as concrete floors, walls, and roofs on site could change the way structures are made as well as increase the speed (no need to stop at 8-hour days) and the quality (computer numerically controlled CNC) and accuracy of structures, where laborers are acting more in a supervisory and material loading capacity than manual construction.

Coordinating Construction

Sequencing, safety, logistics, material storage, deliveries, quality control, equipment management, and reporting, along with a host of other coordination activities, are what make up a construction manager's day on a jobsite. Many aspects of this day-to-day coordination use tools that still operate in silos, requiring that contractors input the same data into multiple tools. Currently, this redundancy is forcing many contractors to focus vendor attention on how tools can communicate information better between their systems to eliminate rework. In his book *Profitable Partnering for Lean Construction* (Wiley-Blackwell, 2004), Thomas Cain estimates that in any given construction project there is "up to 30% waste in labor and materials."

If rework and wastefulness in the construction process accounts for 30 percent waste, the implications for construction consumers are pretty bleak. "Sure, Mrs. Owner, we can build this structure for you. However, we haven't quite figured out how to make our systems work very well with each other so you'll need to pay for these inefficiencies."

This isn't a very compelling offering.

Imagine if Ford, Chevy, or Tesla Motors sold cars on the same premise: "Sure, we can build you a car. The labor and materials to build it, plus our profit, equals X, but because we haven't worked out all the kinks in the way our information is exchanged, you'll need to pay us 30% more." Would you buy the car? You probably will begin looking into public transportation options in detail.

Unfortunately, this condition is true of much of the technology we use to coordinate construction work. The upside is that there is a tremendous opportunity in the analysis and coordination space to create integrated tools that work together in a real-time manner to better inform decisions and to respond to information more effectively.

Another factor is more tools migrating to the cloud. This is creating interesting methods of tapping into existing data that can be repurposed for other uses. In later chapters, we will go into more detail on companies that are beginning to more broadly understand this need in the coordination space and are developing tools that better enable construction managers to use quality information more effectively to deliver better construction products. Overall, the market seems to be consolidating the number of tools and interconnecting those tools used to eliminate waste.

Using Mobile Devices

The use of mobile hardware in construction has changed the landscape of the modern-day project, in both how information is accessed and how it can be added and disseminated. Mobile technology–enabled platforms are now able to communicate between systems and project stakeholders in near real time. One of the major issues of the past in construction has been the inability for feedback and information flow to come from the field fast enough. Now with many of these barriers removed, teams are collaborating using applications (apps) that ride on mobile platforms such as iOS (Apple), Android, Google, or Microsoft Windows (Figure 1.18). These applications are optimized to perform quickly and have created a new dynamic in efficiency, closing the feedback loop.

Figure 1.18 Tablet being used in the field

Additionally, the use of these mobile devices is being defined in different ways. As an example, some teams now employ the use of digital documents or hyperlinked construction document sets. This systems links construction drawings, specifications, submittals, requests for information (RFIs), and other information in PDF form to each other. The value in digital documents is the enhanced visibility into real-time construction information. This all but eliminates versioning control because a team is only able to look at the "latest and greatest" set of information. It is safer to view documents on a tablet in the field because tablet devices can be attached to a harness or safety vest in lieu of large plan and spec sets. Lastly, this example is more sustainable and reduces the amount of printing and the associated reprographic costs.

New costs are associated with mobile-enabled construction, including hardware costs for the devices, application costs (though often much smaller than larger software purchases), wireless or data plans, and staff training. These investments are being made for many construction companies with claims that they are working more efficiently with fewer errors in information exchange and faster response times.

Controlling Schedules

During construction a number of BIM and mobile tools can assist in verifying schedule progress. Usually the key informational needs from contractors during construction are as follows:

- Create and verify the accuracy of the schedules logic
- Introduce detailed sequence schedules to mitigate installation risk
- Manage material and equipment delivery timing
- Manage crew sizes and productivity to align to expected completion timelines
- Use lean scheduling methods to increase productivity between milestones
- Verify work put in place and subcontractors' billing percentages
- Identify root causes that delay or inhibit productivity
- Adjust schedules in real time based on site feedback
- Establish performance incentives/deductions based on scheduled work
- Report to an owner on construction progress and delivery timelines

These are just some of the ways schedules impact the construction process. For many contractors, the golden standard for project schedule controls is Critical Path Method (CPM) scheduling. A significant amount of data points to this type of scheduling method as being largely ineffective in construction. In a recent *SmartMarket* report by McGraw Hill on "Lean in Construction," the results showed that:

- 86 percent of clients are expecting shortened schedules enabled by a lean approach.
- 62 percent of contractors acknowledge that current practices are inefficient.
- 84 percent of construction managers find that adopting lean has led to higher-quality projects.
- 64 percent of construction managers have improved profitability.

Thanks to statistics like these combined with the economic downturn, many contractors have doubled down on their investment to research other means of creating project schedules. Scheduling methods such as location-based scheduling, or "flowline" scheduling, and Q scheduling, or quantitative scheduling, are leading to a renaissance in construction.

One of the biggest innovations in the creation of construction schedules is the ability to collaboratively develop schedules with project stakeholder commitments in the form of pull plans and lean scheduling methods, enabled by technology such as

Pull Plan and Adept software (Figure 1.19). It's interesting to think that Henry Gantt developed the first Gantt bar scheduling method over 100 years ago. If Henry had had better tools, such as data-enabled tablets and real-time applications, do you think he would've developed something different?

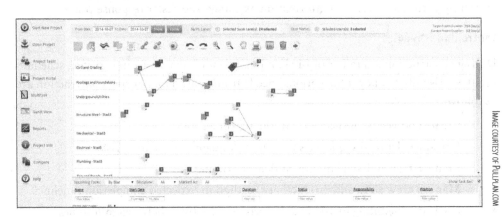

Figure 1.19 Screenshot from Pull Plan, lean planning tool

It's hard to say, but we do know that collaboration trends in construction management will not end with BIM. In fact, the rise in collaborative tools continues to expand to other workflows, including estimating and code checking. Specifically, collaborative scheduling has the potential to become much more than a static multipage Gantt chart. Contractors are beginning to understand the value of completing the elusive feedback loop from the field to inform many inputs to financial data, work planning, and safety metrics. Additionally, they are taking advantage of mobile and BIM-enabled applications that allow them to collaborate with their teams better as well as use schedules in a more meaningful way.

Controlling Cost

Understanding project costs and cash flow is the lifeblood of a project. BIM-derived data can be a valuable source of information from a design model to better inform estimates, reduce assumptions, and create a better dialogue early in a project's start. Yet how do we carry the ability to control costs into the field? Traditional methods included the use of multiple spreadsheets, detailed budgeting breakdowns, and usually some methodology of determining a percentage value of installed work.

As mentioned previously, the industry trend of collaboration and real-time input is continuing into cost controls and is creating a shift in input as data and the ability to report from the field begins to trump input from the office. Think about it. Would knowing on a daily basis the percentage of completion a subcontractor is at be useful to control billing and cash flow? Of course! Additionally, initial BIM-derived estimates created for a project are extending into construction as scope budgets are defined. Some teams are finding ways to continue to connect cost data as construction continues by validating correct billing percentages with custom scope schemas built into tools like Navisworks. Another approach is to use estimating platforms such as Vico and laser scanning to validate the cost of the work put in place versus a visual verification of that in the field (Figure 1.20).

Figure 1.20 Laser scan and BIM overlay

Managing Change

Change in construction is inevitable, or in the words of the ancient Greek philosopher Heraclitus of Ephesus, "Change is the only constant." In construction, each project is unique. Even prototypical projects will face unique site, jurisdictional, and labor nuances from job to job. Because it is unique, there is no "exact science" to the field of construction. This isn't to imply that there are no constants in the way the industry goes about building structures. Rather, BIM and mobile technologies have proven that the fundamental systems to construction can be improved upon with applied innovation.

In 2004, Elon Musk took an active stake in an electric car company called Tesla. Many believed that there simply was no way that Tesla could design and create an all-electric car that competed with large automobile companies, let alone build the infrastructure required to support it. However, over the course of the last decade we have seen that Tesla has not only created a car company that competes with Ford, Chevy, and GM, but that other car manufacturers are looking to partner and learn from Tesla. As of 2014, Tesla is worth $32.4 billion and in many aspects has reinvented how to design and manufacture innovative vehicles.

Manufacturing and construction are similar in some regard. Whereas the activity of starting with parts, available manpower, and equipment is the same, the two quickly travel down divergent paths once field activities begin. Factors include unique site conditions such as weather, permitting, environmental and safety agencies, and other dynamics such as availability of information coming from a design team:

Manufacturing In manufacturing, work commences only after drawings and specifications have been fully engineered, with each part being made to precision in exact quantities that is typically being assembled in a plant. Production is in a controlled environment and efficiencies can be engineered in because of the redundant type of product.

Construction This process is different in construction as often designs are not fully completed and almost always require some degree of field clarifications. The work product is often not redundant and is specialized. Sometimes details are still being worked out, which may affect the schedule, material purchasing decisions, and constructability.

Of course, not every project faces this type of delivery style where the design and construction are concurrent activities. Some projects have the advantage of planning and a thoroughly thought-out design that then carries out to the field. However, in either scenario there will come a point when information is needed and changes need to be discussed and executed in a flexible way.

The way in which information can be managed is where BIM shines. Recent developments in mobile and cloud-based tools now allow us to compare different versions of models to better understand what changed from previous solutions. Additionally, BIM in the field provides a number of significant ways in which it can be leveraged to deliver effective change management strategies that allow teams to analyze potential impacts by using the model for conflict detection, sequencing, and accessibility.

The Closing Construction and Manufacturing Gap

The construction industry is taking more cues from the manufacturing industry, particularly in the area of lean, task management, and flow. Even though there are differences, some of the tools used in manufacturing—such as apps using real-time "line" feedback, prefabrication, 3D printing, and lean methods of communicating—are increasing productivity and changing the way success is measured. In later chapters, we'll discuss the importance of how teams put systems in place to manage the change process in design and construction and the criticality of clarity in which tools are to be used and why response systems are equally important.

Material Management

Material management begins early in the construction process with design. Typically designers specify certain materials and assembly methods. Then the construction team begins to estimate the required quantities for the various parts and pieces of the project and goes to work getting pricing or bids. Procuring the materials is the next step; material orders are placed, or if there is available inventory from the vendor, the contractor will begin to work out delivery to align to the schedule. Depending on the project, some sites have the luxury of material "lay-down" areas. If they don't, materials are delivered on site as needed for the construction work to be done that day or week. In lean terminology, this just-in-time (JIT) approach requires additional coordination, but JIT can achieve a higher degree of accuracy and waste reduction if done correctly. Because of the cost and the importance of having materials on hand to assure the continuous flow of work on a project, it is very important that the construction management team understands what is available and where it is located.

The need to track materials that are coming and going into the project site has been addressed by a number of software vendors such as Prolog Mobile and Zebra. Inventory management applications use a standard barcode, QR code sticker, or RFID tag that communicates with a central database to show the current snapshot of inventory on a site. These systems have pros and cons, and it is typically at the discretion of the project team to determine which technology best addresses their needs. There is new technology in this space that is communicating back to the BIM data to verify installation and shade or layer the geometric component in a unique way, thus updating the virtual model to reflect what is occurring in the field in real time.

Tracking Equipment

Similar to material management, equipment tracking is the ability to understand, at any point in time, the equipment on site, what licensed operators are needed, and what safety records or prework procedures are associated with these tools. Equipment tracking is different from inventory management in that equipment may stay on site for the life of a project and there are unique parameters in operating equipment that extend beyond material tracking. Construction managers can use mobile devices to input and collect information about particular pieces of equipment.

Many equipment tracking applications, such as Verizon's Networkfleet and ToolWatch are available that allow a construction manager to scan a piece of large equipment to access the make and model, load capacities, operations manual, maintenance records, and licensed operators as well as engine diagnostics and fuel consumption. Additionally, personnel can be tracked and managed in the event of a safety issue. One such example is the *hardhat barcoding*—workers scan their helmets into and out of a project site. In the event of an emergency, the construction manager can quickly extract all crew members and take an accurate count of personnel to verify evacuation is complete. Very little in the way of metrics is available yet, but you can easily see how a tool like this could effectively improve a construction manager's safety program.

BIM also plays a role in equipment management by allowing us to design into the model the right equipment to use and its location. For example, a project site may need two cranes working at different heights. Using BIM, you can integrate equipment lift schedules into sequencing simulations to increase effectiveness and the timing of multifloor or multiphase work.

Closeout

Project closeout is the final inspection and submission of a project that includes the facility itself, supporting documentation, and concluding payment per contract documents. Closing out a project is often an afterthought in construction. This is due to a host of reasons—project fatigue, time constraints, or budget constraints. Properly transferring information to an owner at the end of a construction project is rarely done once the building is completed. In fact, in many cases the information is delivered weeks or months after a project has been completed.

BIM has a unique role to play in the process of closeout. As information becomes more integrated with each of the various systems, the ability to transfer linked and more meaningful closeout information is becoming a reality. Here are some examples of information that may be typically included in closeout or turnover:

- Operations and maintenance manuals
- As-built information
- Shop drawings
- Material lists
- Warranty information
- Other information for life-cycle care

Traditionally, this information has been delivered in hardcopy form and/or as electronic files saved to external hard drives. Recently, construction consumers are increasingly requesting information to be delivered digitally. This is creating a unique discussion among the construction industry as to where the "line" of construction stops and operations start, as well as what liability can be associated with delivering an as-built BIM. Largely, the answer has been that owners are starting to see the construction team as the logical steward of their information throughout the construction process in regard to as-built information and other installation records.

The formats being requested are changing as well. Many owners face constricting operations budgets. Because of the need to deliver better turnover data, contractors are using this as a point of differentiation in their businesses to make a more compelling value proposition versus their competitors. Many are doing exciting work in defining what the "as-built" of the future may look like, which includes an artifact and constant component to it that we will cover further in Chapter 7, "BIM and Close Out."

Punch Lists

Punch lists are a checklist of items that must be completed in order for a construction project to be accepted by the owner in order to receive final payment. To get to an as-built, you have to first be able to manage the completion of the work. This means details. The details of punch lists span not being able to start up an HVAC unit (commissioning) to paint chips on the corners of walls. The challenge is in managing the information around punch lists. Applications such as BIM 360 Field and Prolog Mobile allow users to mark up plans on their mobile devices and identify areas that need to be repaired before the contractor's work is marked as completed. Additionally, the use of the BIM model is becoming increasingly useful as these plans can quickly become congested, and thus the use of the model to mark in three dimensions exactly where an issue is can save a significant amount of time (Figure 1.21).

Figure 1.21 Model punch list callout

Construction management teams spend a lot of time resolving punch list items in the field. The efficient closeout of a punch list can allow contractors to get paid faster, turn over buildings earlier, and mitigate the time "wasted" in managing spreadsheets and manual plan markups.

Managing Facilities

Construction consumers continue to become more informed. In fact, the number of owners requiring BIM on over 60 percent of their projects surged from 18 percent in 2009 to 44 percent in 2014 (Figure 1.22). The main driver for owners requiring BIM was reduced errors and rework. Additionally, many owners are now requesting that construction managers deliver as-built information in both hardcopy and digital formats. Some progressive owners are requesting that the information they receive be formatted in a way that can be seamlessly integrated into their computerized maintenance management systems (CMMSs) and/or facility management (FM) systems as part of the project turnover effort. Both CMMSs and FM systems involve managing the operations and assets within a facility effectively to ensure its continued use and can represent up to 85 percent of a facility's total operational costs (Figure 1.22). This trend addresses the risk owners face when given large volumes of hard or static data that needs to be manually input into their maintenance and operational systems.

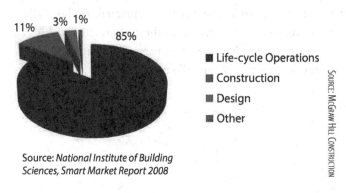

Life-cycle Facility Costs

11% 3% 1%   85%

- Life-cycle Operations
- Construction
- Design
- Other

SOURCE: McGraw Hill Construction

Source: *National Institute of Building Sciences, Smart Market Report 2008*

Figure 1.22 Life-cycle facility costs

From the owner's standpoint, the current turnover process creates significant risk in the form of delay in successful operations. For example, if an owner receives the project turnover data at the end of a project and the first day of use, the facility and operations team's work is just beginning. They now must input this information into their systems to begin operating the structure. This may include changing filters, oiling equipment, replacing damaged equipment. or keeping items under warranty with accurate records. If it takes the facility manager 3–6 months to input this information into a CMMS or FM system, what has been missed? What hasn't been maintained? And what is at risk of falling out of warranty? These are the real issues that face a real estate asset holder in regard to new construction.

To address this knowledge transfer gap, construction managers are becoming more adept and knowledgeable about how to deliver information into the owner's systems at the end of a project. Chapter 7 describes an innovative approach to creating an improved deliverable for a customer.

Knowledge Platform Population

One of the areas growing in importance is the ability for a team to collect and aggregate best practices in a central location that is accessible by a company's associates. These hubs, or knowledge management platforms, are typically web-based tools that allow users to input and search for information about an array of topics. Knowledge management platforms include best practices, innovations, case studies, and tutorials. In many cases, this knowledge base increases the value of a company and the quality of an associate's tenure there, making a company a better place to work.

As it relates to technology and BIM, many companies are uploading video tutorials, which share a user's screen to eliminate the creation of BIM/CAD standards that were traditionally hardcopies. The use of brief video content and knowledge sharing libraries is consistent with the trends of attention span, time commitments, and content absorption related to information finding. Based on metrics from the National Center for Biotechnology Information, US Library of Medicine, the average attention span for a human in 2013 is 8 seconds (Table 1.1).

▶ **Table 1.1** Average attention span statistics for 2013

Attention span statistics	Data
The average attention in 2013	8 seconds
The average attention in 2000	12 seconds
The average attention of a goldfish	9 seconds
Percent of teens who forget major details of close friend and relatives	25%
Percent of people who forget their own birthdays from time to time	7%
Average number of times per hour an office worker checks their e-mail inbox	30
Average length watched of a single Internet video	2.7 minutes

Source: National Center for Biotechnology Information, US Library of Medicine

If viewers believe the information to be valuable, they will dedicate their attention span for longer but rarely for more than 10 minutes total. Because of this reduced amount of focus amid other time commitments throughout a day, we see that many companies have opted to create 5-minute or less videos to describe a goal, workflow, or software function.

Ultimately, knowledge management platforms are used to address many industry challenges, which include:

- An aging workforce, with years of valuable experience and knowledge that can be captured
- New hires' ability to learn and gather information quickly to be up and running faster
- Improvement of a company's overall shareholder value
- Engagement capabilities that allow users to post, share, and refine information in an enterprise

- Reduced commitment in the creation of company standards or process that may change due to new technology

- Faster creation of content as there is less effort in speaking over a video of a live screen environment than typing content

The knowledge capture feature in firms continues to become more valuable as research and development or pilot initiatives are identified and can be used as a means to improve and engage a workplace.

Where the Industry Is Headed

The construction management industry as a whole is headed toward a results-focused toolset that is connected, collaborative in nature, and mobile ready. Although the industry has largely accepted BIM as a useful tool and an improvement to construction management, there are still processes and behaviors beyond the technology that need to be addressed when looking to integrate BIM or to improve on the current condition.

Leadership Buy-In

One of the main issues when BIM first became available to construction managers was the question "Will BIM last?" Of course, since publication of the previous edition of this book the answer is clearly a resounding yes. However, challenges remain for some construction management firms on the leadership side. These challenges are in the form of understanding the value, customer needs, or the costs and return on investment for BIM. This can be due to a number of reasons. Sometimes it's a lack of training on BIM and what it is capable of. Often the ability to explore new tools is limited due to time constraints, where many executives simply don't have the time to check out new tools and processes that may affect them.

With BIM adoption at 70 percent in the United States and growing quickly in other countries per McGraw Hill's "The Business Value of BIM for Construction in Major Global Markets" (p. 8), it is probably safe to say that the majority of firm owners are now aware of BIM compared to five years ago. It is also now becoming an industry norm, whose usefulness is being defined in a number of ways. Just as different firms specialize in building different structures, with different staff and project types, the use of BIM is being applied in different ways that are geared more specifically to those projects. As an example, road and bridge infrastructure projects tend to use less of the functionality geared for vertical construction, such as clash detection and multi-trade coordination. Rather, they tend to use BIM for estimating takeoff, sequencing, cubic yards of cut and fill, and layout, as well as importing laser scans to verify installation accuracy. Another example occurs within vertical construction itself such as the tools required for a school versus a hospital. The needs and complexity of each project type differ significantly.

Because of the unique needs and benefits that BIM and technology can offer a construction management company, it is the responsibility of the team as a whole to define and communicate what they see as valuable and adopt these tools, as well as have a strategy in place to analyze the construction technology space to identify new or improved tools that could further increase their efficiency.

The Evolving Role of the BIM Manager

The role of the *BIM manager* has evolved in construction management. For one, teams are becoming more educated. The number of tech-savvy graduates entering the market who have had advanced exposure or training in BIM tools, combined with current users who are more project experienced, is creating a shift in the use of BIM in construction. Although many firms began with a BIM manager role as a single source of responsibility, we are seeing a shift in dedicated resources to firms now integrating BIM as part of the responsibility of the project team with no dedicated BIM specific staff member. Some projects large enough in size to support a dedicated BIM manager often utilize this resource to the fullest extent. Other firms, whose business model is more diverse, may centralize resources to spread the BIM resource across a variety of projects. Although the strategies of technology and information management in the construction process vary from project to project, details such as project size, budget, and staff availability remain the main factors affecting the decisions and strategy for BIM integration.

At a company level, many construction management firms have shifted BIM-specific personnel into more general technology personnel at a companywide leadership level who consistently seek to understand industry trends and customer needs, identify needed business improvements, and seek out opportunities to increase profits and innovate. There is no hard and fast rule as to where these professionals come from; they often have varied backgrounds. Some come from information technology (IT), project management, software vendors, or BIM. Regardless, construction firms are putting these focused resources in place to manage strategy and deliver technology at a company level as it continues to become more critical to the way construction is delivered.

What Have Been the Results?

The return on investment of BIM has always been somewhat of an elusive metric to capture. However, current data strongly suggests that the introduction of BIM processes and technologies is changing the construction landscape to be more productive with less risk and less in-field changes (Figure 1.23).

Some of the intangibles of BIM and technology are hard to make quantitative. Additional benefits have been increased communication due to more collaborative environments, better information loops from connected applications, virtually connected workspaces, and enhanced capabilities. BIM continues to be built upon as the construction industry finds new ways of using the information available.

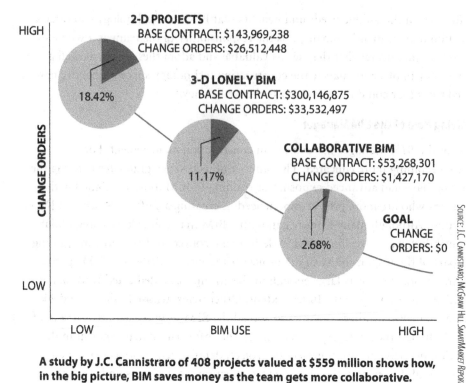

A study by J.C. Cannistraro of 408 projects valued at $559 million shows how, in the big picture, BIM saves money as the team gets more collaborative.

Figure 1.23 BIM-related savings as collaboration increases

Summary

The purpose of this book is to look at BIM and how it continues to change the way we deliver construction. Just as important to BIM as a technology is the journey that BIM has put our industry on. Our industry was in need of reinvention that included technology improvements and enhanced collaboration. Status quo no longer defines a successful contractor. Construction firms around the world continue to realize the benefits of BIM as well as the ancillary features of better methods of sharing and coordinating information.

So is this it?

We don't believe so. In fact, we believe that this is just the beginning for the technological renaissance that is surging within the construction community. Ultimately, the questioning, analysis, and resulting innovations continue to offer exciting opportunities to improve our industry and deliver further on the promise of BIM.

Project Planning

This chapter explores the procedures required to properly plan for BIM on a project. We'll discuss the essential elements that affect the usefulness of BIM in order to establish achievable goals, staff your project, align your team, and create a plan for successful execution of BIM.

In this chapter:
Delivery methods
BIM addenda (contracts)
The fundamental uses of BIM
BIM execution plan

Delivery Methods

The delivery method is the way in which the owner contractually works together with the designer and the builder. It's one of the first decisions an owner makes when deciding to build a structure. According to Barbara J. Jackson (*Design-Build Essentials*, Cengage Learning, 2010), before the fifteenth century this was a relatively simple process: An owner would hire a master builder, who oversaw all aspects of design and construction for an entire project. Over the years, the singular role of the master builder separated into designer (architect) and builder (contractor). It's a general belief that this separation started with Leon Battista Alberti in the mid-fifteenth century. Alberti directed the construction of a new façade on Florence's Gothic church, Santa Maria Novella (Figure 2.1), from plans and models, which, says Jackson, was the "first time in history that plans and diagrams enabled the 'designer' to instruct the 'builder.'" This deviation from the master builder approach is why Alberti is commonly referred to as the first modern-day architect.

PHOTO BY DAVE MCCOOL

Figure 2.1 Santa Maria Novella

Jackson goes on to say that throughout the Industrial Revolution "specialized design and construction expertise was needed to address unique production and facility needs." This specialization created focused efforts within the entities that made up the design and construction teams, which led to the development of professional societies in the mid-nineteenth and and early twentieth centuries. Some familiar ones were the American Institute of Architects (AIA) in 1857 and the Associated General Contractors of America (AGC) in 1918. These societies, by their very name, further segregated the industry. More recently the separation has been exacerbated due to legal requirements such as the Miller Act in 1935 and the Brooks Act in 1972. Both acts led to legal separation between designers and builders. Now the industry is quite fragmented, which makes contracting vehicles more complex for owners.

In order for owners to decide on the appropriate method of delivery, they must answer a number of questions about their project:

What type of building is it?

Which designers have experience on this type of structure?

How risky is this job?

When does this project need to be completed?

What is the budget?

Do we want a contractor on board during design?

Can we legally contract the designer and builder under one agreement?

Can we legally share the risk among the designer, the builder, and ourselves?

The answers to these questions will guide owners in deciding the appropriate delivery method.

There isn't a universal solution, so every project will be a little different. Contractors must understand the advantages and disadvantages of each delivery method to properly implement BIM. In this chapter, we'll consider four delivery methods:

- Design-Bid-Build
- Construction Manager at Risk
- Design-Build
- Integrated Project Delivery

Note: Although this analysis is not an all-inclusive list, it does highlight the more common delivery methods used today.

Design-Bid-Build

Design-Bid-Build (DBB) is the most traditional type of delivery method practiced today. In this method the owner has two contracts: one with the architect and one with the contractor. DBB is considered a linear process because there is no overlap of the architectural services and the contractor services (Figure 2.2).

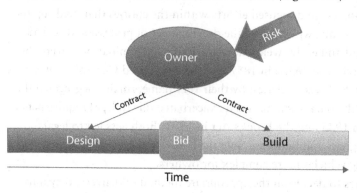

Figure 2.2 Design-Bid-Build

At the beginning of the design phase, the owner selects an architect to develop a program and a conceptual design for the building. The program includes the functional criteria of the building (such as total square footage, type, usage of spaces, performance requirements, number of floors, and building footprint) and is typically in narrative form. The conceptual design could range from hand sketches to a 3D model. Once the owner agrees with the program and conceptual design, engineers (structural, mechanical, electrical, and plumbing) are engaged to further develop the design. The result is a complete set of construction drawings and specifications used for permitting and bidding.

In the bidding phase, general contractors handle the following steps:

1. Review the final drawings and specifications.
2. Collect subcontractor estimates on their relevant scopes of work.
3. Compile the estimates into a complete bid.
4. Deliver the bid as a proposal to the owner.

The owner then awards the project to the general contractor with the lowest bid and releases the contractor to start building the project.

In a perfect world the process is that simple, but in reality a number of challenges may arise. First, it is possible that the low bid is not as low as the owner expected it to be. As mentioned, the architect and engineers create the drawings and specifications with little to no contractor input. This may lead to miscalculations on what is required to construct their design. For example, the architect might have sold the owner on an exterior cladding material that looks amazing. During the design process the owner hires a third party to price the material and it comes in at about $100 a square foot. That's a little more than the owner wanted to pay, but he really likes the look of the material, and he approves it to be specified in the final drawings for bidding. The contractor begins pricing the material to bid the job and discovers that the only place it's manufactured is Italy, the closest person qualified to install it lives in Mexico, and the piece of equipment required to install it is being used in New York.

Furthermore, the material won't be fabricated in time to enclose the building, so temporary enclosures will have to be purchased in order to dry-in the building. To top it all off, the side of the building where this material needs to be installed has a steep slope that won't allow the installation equipment to operate, so special procedures will have to be put in place to allow the equipment to function. The estimated $100 per square foot just went up to $150 per square foot. Due to the increased cost of the project, the owner might not award the project to anyone. However, he now has an idea of where the escalation occurred and can go back to the drawing table with the architect to redesign the building for another round of bidding in the future. Though slightly exaggerated, this scenario is unfortunately very common in DBB and can lead to a slow construction start due to the design-bid-redesign-bid-build process.

Another challenge that may arise in DBB is that gaps or errors are found in the design once the contractor has already begun to build the project. These errors could range from paint colors not being shown in Room 101 to missing a foundation for the column at gridlines A and 1. Regardless how big or small the gap, the contractor will submit a request for information (RFI) to get clarification on the issue. Depending on the complexity of the issue, it could take the contractor some time to compile all the information to accurately describe the problem in a narrative format.

Once the contractor has drafted the RFI, it is then submitted to the owner and architect. If the RFI is as simple as selecting a paint color, the architect can respond quickly and submit the answer back to the contractor. However, if the RFI is complex, the architect may have to involve the engineers in the response. The architect, engineers, and owner will discuss the issue and ideally come to a resolution on how the problem can be solved in a reasonable amount of time, because every day an RFI goes unanswered could delay the construction schedule.

After a resolution is found, the RFI is submitted back to the contractor, who then compiles all costs associated with the added time, material, and labor necessary to make the change. This additional cost is submitted to the owner in the form of a change order, since it is a change to the original design. The owner "owns" the risk associated with these gaps and is responsible for negotiating and paying the contractor any additional costs associated with the change.

Considering the potential difficulties in the Design-Bid-Build method, you may be wondering why an owner would want to use this method at all. For one, humans are creatures of habit. DBB has been the traditional method used for over 50 years, so for many professionals, there's a certain level of comfort in the fact that everyone understands his or her role even though it may not be the most efficient process. Another reason why DBB is still used is due to the false assumption that low-bid procurement is the cheapest method, regardless of the risk of redesign and the change order process. This might make sense for an owner when the building is a cookie-cutter construction—where the program and design have been completely defined and vetted out on multiple construction projects, like a retail chain. Here it might be safe to accept the lowest bid with a high level of confidence, because the only variables are (hopefully) the weather and soil. Finally, owners may prefer DBB because they are able to have absolute control over the design.

Advantages

- The architecture, engineering, and construction (AEC) industry is familiar with this method.
- It is a straightforward competition. If you're the low bid, you get the job.
- It doesn't have legal barriers. This method is accepted in every state and can be used in all markets, including public and federal, provided there are multiple bids.
- The owner keeps a traditional relationship with the architect and has complete control over the design.

Challenges

- There is limited or no communication between the designers and the contractors during the design phase.
- The lack of communication typically leads to cost overruns due to estimates (cost tracking) not being done throughout the design.
- The RFI and change order process can create friction between the architect/ engineers and the contractor, because the gaps or errors have to be justified to the owner who pays for the issues.
- There may be increased litigation due to the lack of collaboration.
- It is a slower delivery method, since the full construction drawings must be completed prior to bid and construction.

BIM in Design-Bid-Build

The DBB delivery method limits the ability for BIM to be used to its full potential, mainly because the builders are not involved during the design phase. This is not to say that builders are the only ones who use BIM fully. It has to do with the way in which BIM is being used. Architects and engineers in DBB use BIM during the design process either due to owner requirements or for their own benefit. They are finding that it is faster to use 3D modeling software than 2D software to produce the necessary documents for bidding.

Note: The value BIM brings compared to traditional drafting is its *parametric* ability. This means that no matter what view you're looking in (plan, elevation, or 3D) the model elements change universally. So, if you moved a door in plan view, it would persist through all the related views (such as elevations, sections, and 3D). This cuts down the time associated with editing 2D views, because BIM creates them automatically. That saves the architect and engineers a lot of time and money.

The 2D bid documents that the architect and engineers produce merely show the "design intent" and do not have to be fully detailed—that is, they don't have to have every detail required for construction. They rely on a process of coordination called "means and methods" in which subcontractors coordinate how the systems will ultimately fit inside the building and submit shop drawings for review and approval. Traditionally this reliance creates subpar drawings because the "design intent" information will ultimately be re-created by the subcontractors. Why would architects and engineers spend the extra effort and money going through every detail when subcontractors have to redo it? The answer is, they won't—and therein lies the problem. Figure 2.3 shows the cost of change during the life cycle of a project in relation to DBB and integrated delivery methods.

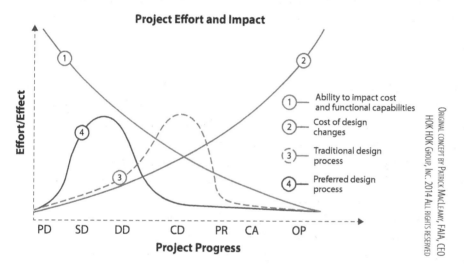

Figure 2.3 MacLeamy Curve

As illustrated in this famous figure, the most beneficial time to solve problems is during the design phase, not during construction. If constructible models aren't produced and/or the personnel involved during design are not involved later to build the project, then you are more likely to miss opportunities to proactively solve issues and save the owner money.

Although BIM may not be able to alter the rigid lines of the DBB approach, it can still add value to this delivery method in the following ways:

- Allows contractor and subcontractors a foundation for coordination of mechanical, electrical, and plumbing (MEP) systems
- Saves subcontractors money by allowing them to prefabricate their systems using the models provided
- Makes the initial estimating process easier for the contractor (depending on the quality)
- Gives another level of clarity to all members on the design and construction of the project, previously only afforded in 2D

The general contractor and subcontractors can find value in utilizing the 3D architectural and engineering (AE) models, if available, for general coordination of systems. The success of this coordination is reliant on the quality of the AE models and the construction experience of the team members participating in coordination. Unless the owner writes it into the contract, the architect and engineers do not have an obligation to share their models. And as stated before, the AE models are typically just showing design intent, so the contractually required documents are still delivered as 2D documentation.

Note: Relying on AE models during 3D coordination can create risk for the construction team without careful review of the details in the construction documents, which will be discussed more in Chapter 4, "BIM and Preconstruction."

Regardless of whether AE models are available, the MEP subcontractors have an obligation to coordinate their systems and produce shop drawings for the engineering team to review. These subcontractors are finding value in parametric 3D modeling, similar to the AE team. By coordinating their systems in 3D, the subcontractors can solve issues early, create constructible models, and prefabricate their systems to save on material and labor costs. For this reason, owners are also starting to require subcontractors to model on DBB projects.

Another value that an AE model brings to the contractor is the ability to generate quick estimates or *rough order of magnitude* (ROM) costs for materials. This process will be discussed in depth in Chapter 4. Again, proceed with caution here.

Note: As a best practice, estimates should always be verified against the contractual drawings in this delivery method.

Lastly, the model is a great visual tool. Many owners enjoy looking at and commenting on their building before it's built. BIM is the modern-day balsa wood or foam board model. It adds an enhanced level of clarity to the 2D documents for the owner to visualize. There's greater visual clarity for other stakeholders as well. The model allows the contractors and subcontractors to complete better constructability reviews with the team, and when the model is combined with a construction schedule, it allows the contractor to analyze the sequence of work for logistics and safety.

Construction Manager at Risk

The Construction Manager at Risk (CMAR) method is similar to DBB. The owner still "owns" the risk of the design and again has to manage two separate contracts: one with the architect and one with the contractor. However, unlike DBB, the contractor is brought in during the design phase, which breaks the linear process of services and is often referred to as "Design-Assist." The owner's agreement with the contractor typically has two parts, referred to as a Part A/Part B agreement:

- The Part A agreement is specifically for the design phase services.
- The Part B agreement is for the construction services.

It is a best practice for the owner to award the Part B agreement to the same contractor that completes the Part A agreement, but owners have the option to bid the Part B services out if they are unsatisfied with the contractor's performance (Figure 2.4).

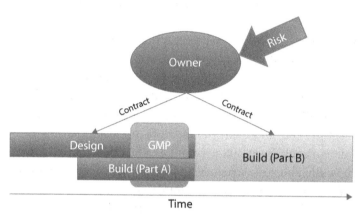

Figure 2.4 Construction Manager at Risk

This method entails a commitment by the contractor to deliver the project within a *guaranteed maximum price* (GMP), which is decided during the Part A, design phase service. The "at risk" relates to the contractor's risk for delivering the building for that GMP. If contractors exceed the price, they lose their fee and profit. That creates a vested interest for contractors in the success of the design and construction. The value of this type of delivery is early collaboration between the contractor and design team during the design process.

The big decision the owner and design team must make is when to bring the contractor on board. Contract drawings progress in three basic phases known as schematic design (SD), design development (DD), and construction drawings (CD). Often, these are broken up into at least two milestones, 50 percent and 100 percent, in order to give the design team a schedule and keep the design progressing. Ideally, owners bring the contractor in somewhere around 100 percent SD or 50 percent DD to make sure there is enough time for them to add value before the construction drawings are complete. If they bring the contractor in too early, the documents may not be complete enough for the contractor to commit to a GMP. This can result in a high contingency or additional money set aside that the contractor allots for the unknowns of the design; that's less than ideal for owners. It becomes a balancing act for the owners, because though they want to have early input from the contractor, they don't want to stifle the creativity of the architect by having the contractor looking over the architect's shoulder with a price book.

Owners also need to decide if they will bring on the contractor or both the contractor and subcontractors. If they choose the latter approach, they will need to specify how the subcontractors are adding value during the design process, since they will be paying a higher premium on the Part A agreement for every subcontractor involved. For example, if the architect's exterior skin design is extremely complex with various materials and glazing systems, the owner might want the contractor to bring on the exterior subcontractors installing that work (glass and glazing, framing and plaster, sheet metal, etc.) to coordinate the details between systems and ensure constructability.

The most apparent advantage of CMAR over DBB is the fact that the contractor can aid in resolving potential issues before construction and can provide estimates as the design progresses. Another advantage with this delivery method is the ability to deliver the project faster. In CMAR, unlike DBB, construction can start prior to all documents being finalized. The building can be segmented into design packages like sitework, structure, core and shell, and interiors, for instance. This means that the structural design could be at 100 percent CD when the interior design is at 50 percent DD, which would allow the structure to get a permit and construction to begin while the architect is still working on the interior design. This works well for projects with complex phasing; it helps reduce the cost of materials due to escalation and speeds up the overall schedule for completion of the building, which reduces the general contractor's overhead costs and allows the owner to start using the facility and generating income.

Advantages

- Contractors are involved early in the project.
- The decision of the contractor doesn't have to be based solely on the price; it can contain quantitative and qualitative factors (discussed in Chapter 3, "How to Market BIM and Win the Project").
- It allows the contractor to run estimates during design and aids in value analysis.
- It allows construction to start before design of the entire building is complete, which can speed up the project.
- The owner still keeps a traditional relationship with the architect by having a separate contract.

Challenges

- The contractor may not be brought on early enough to make a significant impact, which causes frustration with the owner and design team.
- The contractor spends more time competing to win the contract in this delivery, which costs more money (discussed in Chapter 3).

- The owner still has to manage two separate contracts and "owns" all the risk of the design, which means that friction still exists between the design team and construction team.

- Because the owner "owns" the design, the contractor will still be submitting change orders for construction issues not identified during the Part A agreement.

BIM in Construction Manager at Risk

CMAR lends itself well to leveraging all the uses of BIM, as defined later in this chapter. However, the expectations of BIM and the team must be aligned early on by the owner. The owner must create or require a plan to illustrate how BIM will be used, or CMAR faces the same pitfalls as DBB. Let's look at some major reasons for this.

The first issue is caused by AE teams being unable or unwilling to model and/or share their model content. As discussed, the major benefit to CMAR is the early contractor involvement, but the contractor coming on early does not create a contractual obligation for the AE team to model or share their model. Remember, the AE team is typically only required to produce contract drawings, not contract models. It also does not change the mind-set or fee needed for the architect and engineers to coordinate beyond "design intent" if they are modeling. These items must be handled through the contract with the owner, and then the owner must align those requirements with the Part A/Part B agreement to the contractor.

The other issue that could limit BIM's potential is the timing of the contractor's involvement in the design process. For instance, if the owner requires the contractor to use BIM in the Part A services for design constructability review but didn't require the architect and/or engineers to model, the contractor must account for having models created. This might mean creating them in-house or outsourcing to a third party to generate the models, which takes additional time and potentially delays the constructability review.

Second, if the owner doesn't bring in key subcontractors under the Part A agreement, there still wouldn't be constructable models with which to coordinate. The models generated in-house or by the third party would be models derived from the architect's and engineer's design intent, which would be missing the expertise of the subcontractors, therefore limiting the benefit of coordinating with BIM.

Last, if the owner decides to bring the contractor on board later, say at 50 percent CD, there may not be enough time to aggregate models, analyze them for constructability, and adjust the design prior to the architect's and engineer's completion of the construction documents at 100 percent CD. According to the authors of *The Commercial Real Estate Revolution: Nine Transforming Keys to Lowering Costs, Cutting Waste, and Driving Change in a Broken Industry* (Wiley, 2009), "If a contractor is capable of providing full project simulation but is only brought in at the end of the traditional design development phase, most of the decisions have been made." *The threshold for contractors providing BIM services in a CMAR delivery method should be 50 percent DD.*

Owners may have thought requiring BIM from the contractor during design would help reduce the number of gaps and errors found in the traditional DBB method, but if they don't define the expectations of how BIM will be used and align it across the two contracts, they may be very frustrated with the outcome because of the parallel efforts. This method has great potential for BIM use, but it can also contribute to losing an owner's confidence in BIM.

Design-Build

The Design-Build (DB) delivery method is set up to be one of the best methods for promoting collaboration between designers and builders because their services completely overlap and they are required to function as a team. The owner manages just one contract with the design builder and no longer "owns" the risk. The risk is now on the design builder (Figure 2.5).

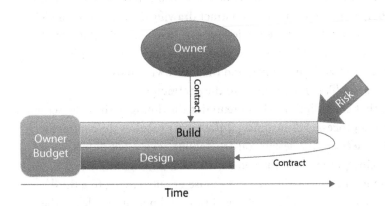

Figure 2.5 Design-Build

In this method, owners should have project requirements and a budget set for how much they want to spend. The owner's project requirements would contain general requirements for function, form, and performance of the building as well as a schedule for completion. The goal of the design builder is to take the performance requirements and "design to the budget" as opposed to reacting to the design and "budgeting to the design." It is the responsibility of the design builder to manage the design process, obtain the necessary permits for construction, and build the project on time and within that budget. The project can be contractor-led, designer-led (architect or engineer), integrated firm-led, or a joint venture. According to McGraw Hill's 2014 *SmartMarket Report* on Project Delivery Systems, Contractor-Led Design-Build is the most popular arrangement in the U.S. and is expected to significantly increase in use over DBB and CMAR by 2017 (Figure 2.6).

Expected Change in Use of Established Delivery Systems in the Industry by 2017
(According to Owners, Architects and Contractors)
Source: McGrow Hill Construction, 2014

Owners ▪ Architects Contractors

Figure 2.6 Expected Change in Use of Established Delivery Systems in the Industry by 2017

Design-Bid-Build

Increase

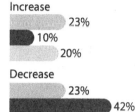

- 23%
- 10%
- 20%

Decrease

- 23%
- 42%
- 30%

Design-Build

Increase

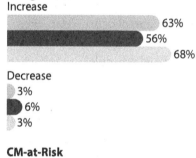

- 63%
- 56%
- 68%

Decrease

- 3%
- 6%
- 3%

CM-at-Risk

Increase

- 50%
- 40%
- 50%

Decrease

- 7%
- 7%
- 4%

Design-Build breaks the mold of traditional roles and responsibilities and demands a collaborative team environment based on trust. This method is a platform for innovation, speed of delivery, increased quality, and budget assurance, but requires a complete sociological change from the project's members to be a success. Some of these changes include:

Worry-Free Owners For owners, the risk is minimized. They no longer have to worry about being held responsible for gaps and/or errors in the design and paying out-of-pocket costs because of these surprises. The risk is now on the shoulders of the design builder, but sometimes this can be a tough change for the owner. The idea of "letting go" of the design and trusting that the designer and contractor are working in their best interest, designing a project that meets or exceeds their expectation, and delivering a high-quality building within their budget can sometimes seem too good to be true. This is particularly true if they're used to driving the process.

Owners Are No Longer Mediators   Another change and relief to owners is that they're no longer the mediator between the AE team and the contractor—managing two entities and separate contracts. This is a great benefit to the collaboration of the team but requires a different mentality. The owner no longer has the direct connection with the architect and has to trust the contractor is not suppressing the creativity of the architect's design to stay within the budget.

Faster Delivery, Faster Decisions   There is one change that affects everyone on the team equally: a faster delivery means faster decisions. This delivery method is set up for the design process to move much faster. Similar to CMAR, the design can be released in increments, but because the design and construction services *completely* overlap, the speed of delivery can be even faster. The comparison in Table 2.1 shows five areas in which Design-Build outperformed DBB and CMAR in cost and schedule.

▶ **Table 2.1** Delivery method results comparison

Metric	Design-Build vs. Design-Bid-Build	Design-Build vs. CMAR
Unit cost	6.1% lower	4.5% lower
Construction speed	12% faster	7% faster
Delivery speed	33.5% faster	23.5% faster
Cost growth	5.2% less	12.6% less
Schedule growth	11.4% less	2.2% less

Construction Industry Institute (CII)/Penn State research comprising 351 projects ranging from 5,000 to 2.5M square feet
The study includes varied project types and sectors.

The owner no longer owns the risk of project delay, so the contractor will take a more active role in driving the design deliverables to meet the construction schedule. That means there will be additional pressure on the owner to engage and describe their future tenants' or user groups' needs in order for them to be incorporated into the design before it's complete. If the owner fails to get the information to the contractor in time for the deliverable, tenants might be unhappy with their spaces or change orders might be introduced due to added scope after the design and budget are finalized. This goes against the entire concept of Design-Build, because this method's ultimate potential is building a project with *no* change orders or additional costs.

Trust and Collaboration   For architects, DB offers a new collaborative approach with the contractor, as opposed to a defensive battle over who is at fault for holes in the design.

However, this is the biggest sociological change, because traditionally there's a mutual lack of trust due to how the contracts are set up in the other delivery methods. Figure 2.7 illustrates what happens when the architect and contractor keep a traditional relationship in a DB delivery.

Figure 2.7 Hole in the boat

There has to be trust and an understanding that the team is in the same boat. Once they realize they're in the boat together, they start becoming more collaborative and innovative, and begin producing higher-quality buildings.

Support Design Creativity Contractors carry a large amount of risk and responsibility, because they are often the lead on the Design-Build team. They also have "boots on the ground" experience in making the architect's design a reality and are able to bring a wealth of lessons learned on materials to the table in order to capitalize on their successes. However, the ownership of risk and knowledge of best practices can sometimes cause the contractor to lead the design instead of supporting the design.

> **Note:** Contractors are not necessarily designers or you'd probably hear conversations like this: "What you need, owner, is a square building with punched windows. You don't want that crazy curtain wall. It's expensive and will create too much heat gain. Oh, that terrazzo stuff, yeah, you want carpet tile or polished concrete."

Contractors know what the most sustainable and economical products are, because they're the ones who have to warrant the work, but if the owner wanted a cookie-cutter building then they'd probably be working under a DBB contract, not DB. The contractor has to support the architect's creativity with lessons learned while maintaining the budget. This is difficult, but a good contractor will find this balance. After all, the top priority of the contractor should be to please the owner with the best building possible.

Hopefully you can see that the DB method requires a considerable amount of team building and alignment. But when team members trust one another and understand their roles, the results can be incredible.

Advantages

- This method encourages collaboration.
- The decision of the contractor doesn't have to be based solely on the price—it can contain quantitative and qualitative decisions (as discussed in Chapter 3).
- This method has the potential to be the fastest delivery method from design to construction.
- The owner manages only one contract.
- The contractor can manage the cost by running estimates throughout the entire design process.
- This method encourages innovation.
- There is the potential for zero change orders.

Challenges

- This method is not traditional and requires a trust-based and collaborative team for success.
- The contractor and architect typically spend more time competing to win the contract in this delivery, which costs more money (as discussed in Chapter 3).
- It is not a widely accepted delivery method in every state for publicly funded projects yet, though federal military and government work is often design-build.

BIM in Design-Build

The Design-Build delivery method opens up the opportunity to fully leverage BIM tools and practices, but the most valuable benefit is having constructible models during design. In other delivery methods, BIM is limited because models have to show only "design intent" for all or the majority of the design process. This limits the ability to solve constructability issues early in the cost curve. With Design-Build, constructible models can evolve throughout the entire design, which serves two purposes:

- It is a leaner process because efforts aren't doubled between engineers and subcontractors.
- The model is constructible, which allows the team to be proactive with issues as opposed to reactive.

The most popular 3D software platforms used for design does not integrate seamlessly and/or meet the needs of the subcontractors who are fabricating from the model yet. This means that architects and engineers might use one platform to create design models (Figure 2.8) and the subcontractors may create a completely separate model to use for fabrication (Figure 2.9). The purpose of the model is essentially the same except that the subcontractor's version is typically more detailed, is coordinated with the other trades, and can be used for fabrication. Wouldn't it be better if there were one model that evolved instead of doubling the effort? That is what can be achieved with Design-Build.

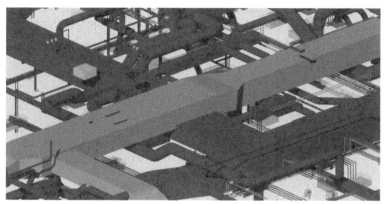

Figure 2.8 Engineer (Autodesk Revit) model showing design intent at 100 percent CD

Figure 2.9 Subcontractor (CAD) fabrication model for shop drawings

Design-Build offers a couple of ways to enable single model evolution:

An Integrated Engineering and Construction Firm   The first method is having an integrated firm that does both engineering and construction. This enables the firm to use a single modeling platform that will allow the engineers to both design and create drawings for permits and then continue to develop the same model into fabrication for construction.

An Engineering Firm and a Subcontractor   The other method is having a joint venture between an engineering firm and a subcontractor. In this method, the engineer dictates and validates the design but the subcontractor performs the modeling. The engineer doesn't model. The subcontractor's model will be "flattened" for the permit drawings and the engineer will stamp them. The subcontractor will then continue to develop the single model into fabrication.

Integrated Project Delivery

The Integrated Project Delivery (IPD) method is similar to Design-Build. However, the major difference in this method is the risk (Figure 2.10). In DBB and CMAR, the owner "owned" the risk. In DB, the design builder owned the risk. In IPD, the risk is shared between the owner, architect, and contractor, but so is the reward. The contract becomes a multiparty agreement.

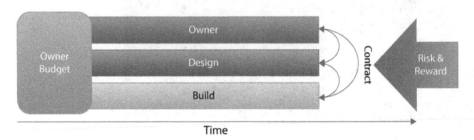

Figure 2.10 Integrated Project Delivery method

IPD promotes the concept that by sharing the risk and reward of a project through target project goals, compensation may increase or decrease depending on results. Say, for example, the team, including the owner, develops a goal for the entire project budget. If the project comes in under budget, additional fees are distributed to the team; if the project comes in over budget, fees are reduced. By holding all members accountable, IPD fosters a great degree of communication and promotes intense collaboration among the project team, because it can result in additional profits for all members of the team.

Conceptually, these incentives can help teams overcome the sociological challenges found in Design-Build. Now everyone's in the boat, but everyone is incentivized to recognize whether there's a hole and work together to fix it instead of working for individual fees. IPD also works another way by rewarding innovation as opposed to just encouraging innovation. Now the entire team wants to create value in selecting the best materials and innovating the design. In this model, the team may decide that a boat isn't the right vehicle at all and another solution may work better.

Advantages

- This method incentivizes collaboration.
- The decision of the contractor doesn't have to be based solely on the price—it can contain quantitative and qualitative decisions (as discussed in Chapter 3).
- This method has the potential to be the fastest delivery method from design to construction.

- The contractor can manage the cost by running estimates throughout the entire design process.
- This method incentivizes innovation.
- There is the potential for zero change orders.

Challenges

- This method is not traditional and requires a trust-based and collaborative team to succeed.
- The contractor and architect typically spend more time competing to win the contract in this delivery, which costs more money (as discussed in Chapter 3).
- This method is not accepted by any public agencies yet.

Note: BIM in Integrated Project Delivery, similar to Design-Build, presents the same opportunities to fully leverage BIM. However, the incentivized program may create a more innovative team—a group open to research and development (R&D) and thinking outside the comfort zone of proven tools.

BIM Addenda (Contracts)

The McGraw-Hill *Multi-Year Trend Analysis* from 2007 to 2012 showed that BIM adoption in North America grew from 28 percent in 2007 to 71 percent in 2012. This "BIM boom" was a global trend, and as it started to grow in popularity, many questions were raised among AEC members. Dwight Larson and Kate Golden define the various questions raised by the industry in their article "Entering the Brave, New World: An Introduction to Contracting for Building Information Modeling" (William Mitchell, *Law Review* [34:1], available at http://open.wmitchell.edu/cgi/viewcontent .cgi?article=1234&context=wmlr):

> *Does it alter the traditional allocation of responsibility and liability exposure among owners, designers, contractors, and suppliers? What are the risks of sharing digital models with other parties? Does the party managing the modeling process assume any additional liability exposure? What risks arise from potential interoperability of the various BIM software platforms in use? How should intellectual property rights be addressed? What risks arise for the party taking responsibility for establishing and maintaining the networked file-sharing site used as a depository for models? How might BIM alter the set of post-construction deliverables on a project, and what are the implications of the changes? And, perhaps most importantly, how can the project contracts enhance rather than limit the benefits to be gained through the use of BIM?*

In response to these questions, a number of professional societies began to develop addenda to their existing contracts to protect a party's risk, standardize execution, and define responsibilities of team members on projects with BIM requirements. An *addendum* is an additional obligation that is outside the main contract. The three major agencies involved in creating these addenda were the American Institute of Architects (AIA), Associated General Contractors of America (AGC), and the Design-Build Institute of America (DBIA).

These addenda provide owners, architects, and contractors with a template for developing a project-specific BIM addendum (Table 2.2). However, this requires that the parties entering the agreement have a thorough understanding of model level of development (LOD), model uses, model sharing, and ownership privileges. Although this is fine for an experienced BIM team, it can be difficult for a team new to BIM to spell out at the beginning of a project what challenges they are anticipating.

▶ **Table 2.2** Addenda comparison chart

Contract	AIA E202	ConsensusDocs 301	DBIA E-BIMWD	AIA E203
Created	2008	2008	2010	2013
Default Lead	Architect	Owner	Design-Builder	Architect
General Provisions	No	Yes	Yes	Yes
Standard Definitions	Yes	Yes	Yes	Yes
Information Manager Roles and Responsibilities	Yes	Yes	Yes	Yes
LOD Matrix	Yes	No	No	Yes
Intellectual Property Rights	No	Yes	Yes	Yes
Model Reliance Risk	No	Yes	No	Yes
Information Exchange Protocols	No	Yes	No	Yes (G201)
BIM Plan Protocol	No	Yes	No	Yes (G202)

It is best for users to either consult with a professional peer who has entered into a similar agreement or bring on a BIM project consultant to help the parties define the best way to distribute roles and responsibilities among the project teams before entering a BIM contract. This approach will streamline the process significantly and can provide valuable insight to avoid potential pitfalls. If neither of these options is available, consult with your legal counsel about the contract language, with a focus on integrating BIM and the project team and clearly defining these roles. As this book will show, BIM is most effective when used as part of an integrated effort among all members of the project team. This is particularly evident during project planning when it is being determined when and how BIM is to be used, shared, and analyzed.

 Note: Each agency has their own BIM-specific addendum, so we'll focus on those addenda in order to highlight the consistencies and inconsistencies of each one.

AIA: Document E202

The E202 was created by the AIA in 2008 and is one of the first addendum templates released for BIM. It is a simple, architect-focused document broken up into four sections: general provisions, protocol, level of development, and model elements.

> **Note:** The "Level of Development" (LOD) and "Model Elements" sections make up what is known as the "LOD Matrix" shown in Figure 2.12.

This document is arguably the first version of the current BIM execution plan, which we'll discuss later in this chapter. Its simplicity has allowed the AEC industry to grasp some key aspects that need to be defined when using BIM on a project. However, this document is deficient in describing intellectual property rights.

AGC: ConsensusDocs 301

Around the same time the AIA was developing the E202, a number of other entities—including owners, architects, engineers, general contractors, subcontractors, manufacturers, and lawyers—were developing the ConsensusDocs 301. This document was created to be a broad consensus agreement between multiple parties on projects using BIM technologies as opposed to the AIA E202, which was developed and written with a singular emphasis on the architect, stating "architect shall _____" throughout the document.

The ConsensusDocs 301 has a format similar to that of the E202, but its contractual language and protections are more robust. Also, because this document is a consensus agreement, it encourages more of a team approach to leverage BIM technology with responsible data sharing and collaboration to minimize the chances of finger-pointing and/or litigation. Furthermore, having the owner as the default lead creates a neutral party to manage the delegations. The document is broken down into six sections: general principles, definitions, information management, BIM execution plan, risk allocation, and intellectual property rights. This document has more contractual protection than the E202, but it lacks the LOD matrix.

You can find the ConsensusDocs 301 on the AGC's website (www.agc.org/cs/contracts).

DBIA: Document E-BIMWD

The E-BIMWD is designed specifically for Design-Build projects and should be used in conjunction with the other Design-Build contracts. The E-BIMWD is based on the E202 and ConsensusDocs, so it has a similar format. However, it may seem as if it has several gaps compared to the ConsensusDocs, but this is intentional. In a phone conversation with Robynne Thaxton Parkinson, who created the first draft of this document, she stated: "The design-build philosophy when it comes to contracts is less is more," meaning that there shouldn't be redundant language in the BIM addendum and the actual governing contract. The E-BIMWD is supposed to be a "guided conversation" to help the team decide on the uses as opposed to prescribing the uses.

The E-BIMWD addresses how the addendum is used with the other DB contracts, discusses major concerns for sharing intellectual property, and briefly describes the information manager's responsibility to manage files during the BIM process.

This document is available upon request. For information on requesting a copy, visit https://www.dbia.org/resource-center/Pages/Contracts.aspx.

AIA: Document E203

The AIA E203, released in 2013, is a complete overhaul of the E202 in its format and content.

The format changes made in the E203 are the separation of the contractual obligations that stay consistent on projects (general provisions, definitions, and roles and responsibilities) and the project-specific items that are modified on every project (project personnel, model origin, information exchange). This was achieved by adding two documents used in conjunction with the E203 Exhibit. These two documents are called the G201 – 2013, Project Digital Data Protocol Form and the G202 – 2013, Project Building Information Modeling Protocol Form. The purpose for this separation is to allow teams to identify the global expectations of BIM at the outset of a project with the Exhibit E203. This gives the team members a general understanding of what is expected of them prior to being brought onto the job, whether that's for determining the appropriate experience required, pricing scope of work, or both. Once team members are brought on board, the specific approach for information exchange and BIM protocols will be defined and agreed upon through the G201 and G202, similar to a BIM execution plan (discussed later in this chapter).

The general provisions now provide a section that allows team members to notify the other participants if the G201 and/or G202 added additional scope to their services. This is an important clause, because the uses for BIM are not fully defined when the initial E203 Exhibit is distributed. Also, an additional section was added to further define the transmission and ownership of digital data, which was only briefly mentioned in the E202. Lastly, language about the risk of relying on the model was added to specify whether or not model content shared by one participant can be relied on by another.

The E203, G201, and G202 documents can be found on the AIA's website (http://www.aia.org/aiaucmp/groups/aia/documents/pdf/aiab099084.pdf).

Contracts Summary

Looking at Table 2.2, you might assume that the AIA E203 is the best addendum to use because it has "yes" all the way down the columns. It's important to understand that any of these addenda can be used. They're a starting point, and depending on the delivery method, experience of the team, and the desired uses of BIM, each is relevant and has its place. The intention of an addendum is not to point fingers if something goes wrong but rather to clearly define the tasks, responsibilities, and rights at the onset of a project. Unless contract language is defined before creating, using, or transferring BIM technologies into a project, no team members can be held responsible for delivering on their intended goals.

Collaboration, Not Litigation

What's unique about these associations, contracts, and the industry as a whole is the desire to achieve the following:

- Eliminate litigation
- Responsibly use technology
- Create a more collaborative environment

BIM technology requires a change in the process of construction in order to work, and this change must extend into the way we write and negotiate contracts. At some point, our industry turned from a necessary collaboration of construction professionals to its current litigious arena. We can begin changing by looking at how we worked in the past.

Sy Hardin, a structural engineer with 40 years' experience, who was previously the structural department lead for Sverdrup & Parcel says, "The problem with the construction industry is the focus has shifted to lawsuits and litigation prevention as opposed to individual craft. The ability to effectively communicate, in some respects, is more important than what is drawn. Technology should always and without exception better our ability to communicate, not complicate it." Although BIM offers an effective array of tools, it must be built on a strong foundation of information sharing.

The Fundamental Uses of BIM

BIM tools are technology based, which means that every day another software, app, add-in, or cloud-based solution will be introduced into our industry. This can make it difficult for owners and AEC firms to prescribe what tools to use for a project, how they will be used, and who will be using them in the BIM addendum. It is best to first grasp the idea of model development and how it relates to the five fundamental uses of BIM in order to select the appropriate tools and personnel required for the job and eliminate confusion and inefficiencies.

Similar to selecting a crane for a job, you must first understand how tall the building is, the site limitations, and the construction type in order to select the crane. The cranes may get nicer, but the building construction and analysis for selecting the crane doesn't change. After you select the crane, you decide on the certified operator—you don't put the "new guy" on it. The same concept applies to BIM tools and uses.

Level of Development

The majority of people, regardless of profession, understand the concept of a *virtual model*, but what has happened in the industry is that simply stating "virtual modeling" as a requirement has created confusion and frustration. It would be similar to an owner saying she wants "a building" as a requirement, which leaves a lot of room for interpretation. For this reason, various organizations have come up with ways to define the levels of virtual modeling, commonly referred to as the *level of development* (LOD).

The LOD is defined by the AIA E203 as "the minimum dimensional, spatial, quantitative, qualitative, and other data included in a Model Element to support the Authorized Uses associated with such LOD." Simply put, it defines the precision of the 3D elements and the amount of information contained within each element. Currently three major LOD matrixes are being used in the United States: AIA, BIMForum, and U.S. Army Corps of Engineers (USACE). Each one has a slightly different approach, but all of them are based on the same concept. Table 2.3, based on the BIMForum LOD matrix, demonstrates how LOD relates to the quality of the model content.

▶ **Table 2.3** Levels of development

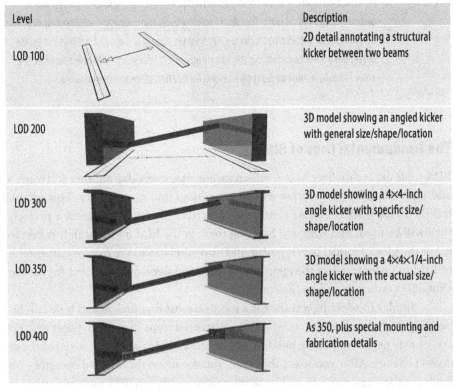

Level		Description
LOD 100		2D detail annotating a structural kicker between two beams
LOD 200		3D model showing an angled kicker with general size/shape/location
LOD 300		3D model showing a 4×4-inch angle kicker with specific size/shape/location
LOD 350		3D model showing a 4×4×1/4-inch angle kicker with the actual size/shape/location
LOD 400		As 350, plus special mounting and fabrication details

As you can see, the idea is to define model evolution. When used correctly, the model adapts; it becomes more useful and smarter; and it's able to survive longer in the building process. It doesn't matter which LOD is used; the important part is that it is used. As we discuss the five fundamental uses, it will become apparent why that is so important.

Model-Based Coordination

Model-based coordination is the foundational use of BIM and where the process begins. Traditionally, there are two essential coordination processes in a project; document coordination and installation coordination. Without good document and installation coordination, the resulting structure is usually of subpar quality with numerous issues during construction as well as during operations. By leveraging BIM, these two coordination processes can be aligned, but doing so requires an understanding of the LOD and alignment of the team.

Contract documents used for permitting and bidding can be produced four ways: by hand, by using 2D software, by using a combination of 2D documentation on top of 3D model views, or by creating exact 3D models with embedded data. To better understand this, let's look at the process of detailing a structural kicker.

- A structural engineer could draw the beams and kickers at LOD 100 in plan view (similar to the example shown in Table 2.3) and another LOD 100 detail in elevation (see Figure 2.11).

- The structural engineer could use a hybrid approach where he models the beams at LOD 200 or 300, but uses LOD 100 to annotate the locations of the kickers in plan view and the LOD 100 detail in elevation (see Figure 2.11).

- The structural engineer could model both the beams and the kickers, but place the kickers in the correct location but at a general angle (LOD 200). The LOD 100 detail in elevation would still be required to complete the design intent (see Figure 2.11).

- The structural engineer could model both the beams and kickers at LOD 300 (or better) and use the parametric capabilities of BIM to automatically generate an elevation view equivalent to that shown in Figure 2.11.

If you looked at any of these scenarios from a bird's-eye view (plan view), it wouldn't make much of a difference for design intent. So every combination produces the same results when they're "flattened" onto paper for *document coordination*, but what the first three scenarios are neglecting is the *installation coordination*.

Installation coordination is accomplished with these steps:

1. Combining the trade models into one consolidated (or federated) model
2. Running a clash detection on the elements or systems within that model
3. Resolving any conflicts between the selected systems (this and other coordination types will be further detailed in Chapter 5, "BIM and Construction")

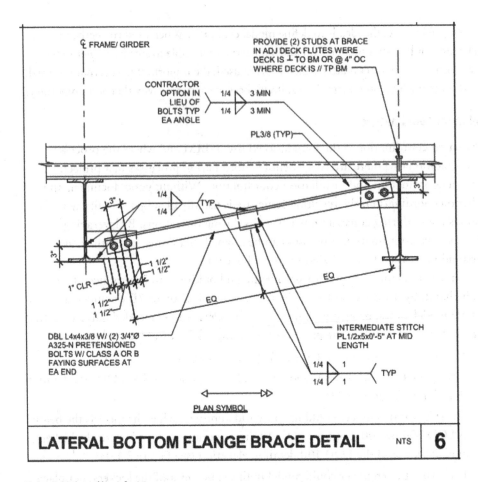

Figure 2.11 Lateral brace frame

It's fairly self-explanatory, but if an element isn't shown in 3D it won't show up as a clash with another 3D object.

The Dangers of Undefined LOD

Let's look at a scenario to better understand the LOD concept with document coordination and installation coordination.

The construction team receives a model from the structural engineer after being awarded the job, but the team doesn't know what the level of development is. "Well, a model's better than no model," the project manager says. He tells the intern to use the structural engineer's model for coordination of the subcontractor's systems. The intern combines the structural engineer's model with the mechanical subcontractor's model, runs a clash test, and *zero* clashes show up. The mechanical detailer is excited about that number, because he worked very hard to keep his systems above the 12-foot-high ceilings by using the beam pockets instead of routing the duct under the beams. After confirming the clash test, the mechanical detailer provides shop drawings to the mechanical engineer to review and approve. The shop drawings are approved and installation begins in the field.

As the steel is being erected, the mechanical subcontractor comes into the jobsite trailer and asks, "What are those kickers for?" "Kickers!" the intern exclaims. The intern goes out to the field and realizes that there are kickers in the beam pockets where the mechanical contractor had routed his duct. He goes back to the trailer to look at the model. "There's nothing shown," he says. He decides to open up the permit set of structural drawings sitting under his desk that have now collected a nice layer of red dirt from the jobsite. He flips to a plan view and notices that there's a 2D annotation for kickers in multiple places on the floor and it references Figure 2.11. His stomach turns into knots. The structural engineer had used a hybrid approach to create the construction documents; modeling major structural components but annotating the kickers at LOD 100. The mechanical duct has no option but to be lowered due to the kickers, so an RFI is drafted and an unhappy owner is faced with lowering the beautiful 12-foot ceilings to 9 feet.

Unnecessary field changes are quite common when implementing BIM without defining LODs or assigning the proper personnel to operate the tools. In the sidebar "The Dangers of Undefined LOD," who is at fault? The owner had "virtual modeling" in the requirements for design and construction but didn't define LOD. The project manager relied on the structural engineer's model and assigned an inexperienced intern to run coordination. The mechanical subcontractor didn't review the structural drawings to route his duct. The mechanical engineer approved the drawings. A number of people could be at fault, but in the end everyone loses. The more BIM is used and relied on for drawings and system coordination, the more important it will be to clearly define expectations and align the LOD requirements during the design and construction phases (see Figure 2.12 for a sample LOD matrix).

LOD Matrix		
LOD 100	Generic representation.	AR Architect
LOD 200	LOD 100 + General size/shape/location	SE Structural Engineer / ME Mechanical Engineer
LOD 300	LOD 200 + Specific size/shape/location	EE Electrical Engineer
LOD 350	LOD 300 + Actual size/shape/location	PE Plumbing Engineer
LOD 400	LOD 350 + Special mounting and fabrication details	FP Fire Protection / GC General Contractor
LOD 500	Field verified installation	TC Trade Contractor

Model Elements Utilizing CSI Uniformat™		Level of Development (LOD), Model Element Author (MEA)								
		Schematic Design		Design Development		Construction Docs		Shop Drawings		As-Built Docs	
		LOD	MEA	LOD	MEA	LOD	MEA	LOD	MEA	LOD	MEA
A SUBSTRUCTURE	A10 Foundations	200	SE	300	SE	350	SE	400	TC	500	TC
	A20 Subgrade Enclosures	200	SE	300	SE	350	SE	400	TC	500	TC
	A40 Slabs-on-Grade	200	SE	300	SE	350	SE	400	TC	500	TC
B SHELL	B10 Superstructure	200	SE	300	TC	350	TC	400	TC	500	TC
C INTERIORS	C10 Interior Construction	200	AR	200	AR	350	AR	400	TC	500	TC
	C20 Interior Finishes	200	AR	100	AR	200	AR	300	AR	500	AR
D SERVICES		200		300		350		400		500	
	D20 Plumbing	100	PE	300	TC	350	TC	400	TC	500	TC
	D30 HVAC	100	ME	300	TC	350	TC	400	TC	500	TC
	D40 Fire Protection	100	FP	300	FP	350	FP	400	FP	500	FP
	D50 Electrical	100	EE	300	TC	350	TC	400	TC	500	TC
	D60 Communications	100	EE	200	TC	350	TC	400	TC	500	TC

Figure 2.12 LOD matrix

Model-Based Scheduling

Another common use for BIM is tying the model (design or construction) to a schedule to animate the sequence of work and display where a project should be at any given time. This has become a common tool for selling work to owners, looking at logistics of construction for efficiencies and safety, as well as throughout construction for justifying subcontractor billings to the owner for completed work. The reason this has become such a popular feature of BIM is its ability to give immediate clarity to all stakeholders in understanding the project schedule. Typical Gantt chart schedules can be hard to understand, but when you watch a simulation of the building being built, the logic becomes more tangible.

LOD plays a role in this use depending on the scheduling methodology (Critical Path Method, Line of Balance, etc.), but the AIA E203 basically says that you can have more precise model-based schedules as the model becomes more precise. While, that's right: The more developed the model, the more detailed your schedule simulation can be. You could show the bolts going in if you wanted to. However, the potential return on the time investment required to link model elements to the schedule diminishes at a certain level of detail.

Model-based scheduling can be used at all stages of the project, whether it's during conceptual design to discuss site logistics or used during construction for demonstrating the sequence of work and validating costs of completed work. It doesn't require a high skill set to create model-based simulations, but it does require a competent person who understands the sequence of construction and scheduling logic. For this reason, it is important to integrate personnel who can achieve both into the project team.

Model-Based Estimating

Model-based estimating has been redefined over the past five years. Originally, it meant that the model-based schedule had costs associated with the elements so it could track time and cost with the animation. The owner would be able to know the exact amount the contractor should be billing at any given time, provided the subcontractor's scope was quantified in the model. Although this method can still be used, more companies use model-based estimating in the form of a *takeoff*. This is where you use the model to extract quantities of materials and associate costs with those materials for estimating purposes.

Using the model for cost estimating should be handled with the same caution as model-based coordination. The LOD has to be defined before you can rely on the model for accurate estimates. You can use the same scenario we considered earlier with the kicker example. If the structural engineer delivered the hybrid LOD model and you used it to extract quantities of steel, your estimate would be off by the cost of every kicker required on the project if the model was the sole source of information. Similar to model-based coordination, experienced personnel in both the technology and in estimating are required to achieve success with this kind of estimating.

Model-Based Facilities Management

Model-based facilities management focuses on leveraging the model information to reduce the owner's costs over the life cycle of a structure (see Figure 2.13). A common figure that is used to describe life-cycle cost of a building states that around 20 percent of life-cycle cost is designing and building and 80 percent is operating and maintaining.

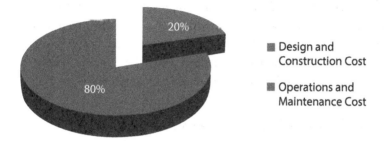

Figure 2.13 Life-cycle cost of a building

That means that there is a high level of importance in making the structures as efficient as possible when it comes to operations and maintenance. Hypothetically, in the kicker example, information could have been added to the actual weld element, shown in the LOD 400 example in Table 2.3, describing it as a fillet weld that needs to be done by a certified welder using E70XX electrodes and that should conform to the latest revision of the American Welding Society's D1.1. Additionally, hyperlinks could have been added to approved websites that sell equipment and to a YouTube video of a worker creating a fillet weld.

The model has unlimited potential to contain all of that information. Apply that same scenario to a light fixture, though. A building can contain hundreds of different light fixtures that are constantly being replaced throughout the life cycle of a building. If the fixture in the model had the serial number, manufacturer, warranty information, bulb requirements, instructions for ordering the bulb, and a video of someone replacing the bulb, then it might not seem so ridiculous, mainly because it is applicable. The information contained within a model can be helpful in reducing costs associated with maintenance and is less likely to be misplaced, deteriorate, or get coffee spilled on it like paper documents. (These concepts will be covered more in Chapters 6 and 7.)

Note: There's another important lesson to be learned in this example on LOD. In the "Caveats" section of the BIMForum's "Level of Development Specification," they state that "there is no strict correspondence between LOD's and design phases. Building systems are developed at different rates through the design process. For example, design of the structural system is usually well ahead of the design of interior construction." This means that the structural steel model could be at LOD 400 the same time the architectural model is at LOD 300. They go on to say that "there is no such thing as an 'LOD _____ model.'" In other words, the architectural model could have doors at LOD 200, ceilings at LOD 300, and walls at LOD 350.

Can you imagine how cumbersome it would be for architects, engineers, and subcontractors if they had to put all that information into every element in the model? LOD can be misinterpreted this way, and when it is, it becomes ridiculous. For this reason, it is important to understand the purpose of an LOD matrix. It allows the owner to define and organize what elements require a certain level of information for use in the life cycle of the building at the onset of a project with the BIM addendum. Once these elements are identified, it is necessary for the AEC firms to align with one another and collaborate with the facilities director on how he or she will leverage that information for maintenance and operations of the building. This collaboration is necessary to make sure that the information contained within the model and related linked database are delivered in the right format to communicate with the building management software.

Note: Model-based facilities management is not a click-of-the-button process, nor is it typical of most projects, so it requires a high level of expertise. It's a best practice to get the team involved early on to set clear goals for how the model will inform facilities management, and may require a third-party expert to assist in outlining the procedure in the contract or execution plan (discussed later in this chapter). Writing "all trades are to deliver LOD 400 models" or "the model will be used for facilities management" in a contract is just as open-ended as writing "virtual modeling" as a requirement.

Model-Based Analysis

Model-based facilities management is often driven by the financial cost to the owner, whereas model-based analysis has more to do with the qualitative costs to the tenants and the environmental costs to our planet. There are some aspects of this analysis that will reduce financial costs to the owner in regard to operation fees, but sustainable designs aren't always the most maintenance-friendly solutions.

For example, windows are proven to promote better production and health among tenants through natural daylight (qualitative costs). In order to have more daylight, you have to have more windows. If you have more windows, there will be more window-cleaning costs (financial costs). The idea, however, is that if the building is designed and analyzed properly, the operational savings far exceed the cost of maintenance for a few extra windows. BIM can play a key role in this analysis.

During conceptual-schematic design, an architect can create simple models at LOD 200 to analyze the location of the site, orientation of the building for heat gain, wind direction for natural ventilation, and the shape for daylighting indoor spaces. Once the conceptual design is vetted out, the model can then be further enhanced with information for material selection, mechanical systems, and electrical systems, which often push the LOD to 300–350. This information is required to analyze loads and project how much energy will be required for a building to operate and keep the tenants comfortable.

That analysis is an ebb and flow between the designs of the architect, electrical engineer, and mechanical engineer. Bigger windows mean more natural lighting and less artificial lighting (thus reducing electrical loads), but bigger windows mean more heat gain from sunlight and possibly more air conditioning (thus increasing mechanical loads). A separate energy analysis software program is often necessary when doing this more detailed analysis.

To fully leverage this analysis, you must ensure seamless communication among the members of the AE team before the start of conceptual-schematic design. An expert with the analytical tools is needed to coordinate the entire process. The need for this analysis is obvious, but often not enough trained professionals are brought on early in the process to realize the full potential of qualitative and environmental cost savings. The governing authorities or owner have to prescribe this kind of analysis if it is going to be used on a project.

BIM Execution Plan

Now that the uses of BIM have been detailed, you can begin to develop an execution plan for the successful implementation of BIM on your project. At this point, you should know what delivery method is being used and how it impacts the added value of BIM, the importance of the addendum in mitigating the risks associated with the uses of BIM, and what capabilities are required from the personnel and software to successfully implement the five fundamental uses of BIM. You've got all the pieces, but how do you execute your plan? That's where the BIM execution plan comes in.

The BIM execution plan is a document that is outside the governing contracts (DBB, CMAR, DB, and IPD) and the BIM addenda. It contains the instructions for executing BIM and is based on the goals the owner or team defined within the contract or addendum. It defines how the team and software will communicate, what the expectations of the team are, and how the information is organized.

History of the BIM Execution Plan

Around 2007, multiple parties were developing BIM execution plan templates. The two most notable templates were the *Penn State BIM Project Execution Planning Guide* and the *Autodesk Communication Specification*. These templates were created at the beginning of the BIM boom, when there was a void in the industry for a project-specific document to help teams collaborate, discuss the goals for the project, define the uses of BIM to be applied, and create a plan of attack for implementing them. A contract and/or addendum could be generated in a silo prescribing every aspect of the use of BIM on the project, but at the time there wasn't an instructional document describing exactly what was required from the team to successfully apply the uses of BIM. As a result, most firms did their own thing because there was inconsistency in achieving the desired results. The construction industry came to the realization that BIM execution plans were necessary in order for BIM to be a success.

One of the most important parts of these templates is the Goals and Uses/Objectives chart. I mentioned in the introduction that goals and uses should be identified in the contract or addendum, but there must also be a way for owners and team members to think through the questions, "What are we trying to achieve with BIM and what is required to achieve it?" before putting a certain use in the contract. Every team member has a different answer to those questions, because each member has a different interest and uses information differently throughout design and construction.

In Figure 2.14 you can see these various goals. The designer wants to know things like aesthetics and code compliance, the contractor wants to know counts for pricing, and the facilities manager wants to know where to order new hinges.

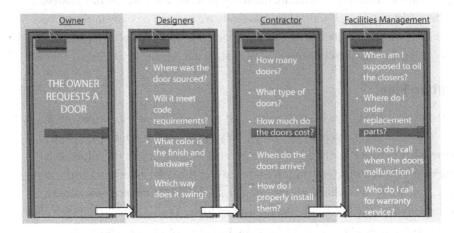

Figure 2.14 Life-cycle information for a door

As we've discussed, huge amounts of information can be embedded or linked to BIM, so how do you decide which goals are more important and align the team? That's what the Goals and Use/Objectives chart aims to do. This chart, shown in Table 2.4, is a basic spreadsheet in both the Penn State and Autodesk templates, but it guides the discussion of the team to align the expectations of BIM for a project. In so doing, the team is able to "begin with the end in mind," which gives direction and focus on what's important to achieving the desired outcomes. Once the goals and uses have been ranked and selected, they can be inserted into the contract or BIM addendum.

▶ **Table 2.4** Goals and Use/Objectives chart

Priority (1–3)	Goal: What is the value?	Required BIM use
1	More efficient installation of MEP systems	Model-based coordination
3	Track billing and schedule with model.	Model-based scheduling
2	Use the model after construction for facilities management.	Model-based facilities management
2	Do cost trending during design with the model.	Model-based estimating
1	Do a computational fluid dynamics (CFD) analysis with the model for natural ventilation study.	Model-based analysis

Based on the Penn State Project Execution Planning Guide

The Penn State and Autodesk templates have shaped the industry standards for BIM planning. Most BIM specifications or owner standards will reference certain sections of these guides. The rest of this chapter is a roadmap to developing a BIM execution plan.

If your company has not yet developed an execution plan, these documents are a good starting point:

- Penn State BIM Project Execution Planning Guide: `http://bim.psu.edu`
- Autodesk Communication Specification: `http://www.thecadstore.com/pdf/autodesk_communication_specification.pdf`

Communication

Seamless communication is the first priority of the execution plan. If all members understand this, the expectation and organization will fall right into place. This one aspect can make or break the efficiencies of a project. You can define the goals, determine all the uses, and have the most experienced BIM team in the world, but a breakdown in communication—verbally among people or virtually in the software environment—will destroy a project.

People

The next time you're in an airport, look around at the people waiting for their flights. You'll notice that technology has indirectly handicapped two of our most powerful senses of communication and created an antisocial behavior. Most people are now fixated visually and aurally to their laptops or mobile devices with complete disregard for their surroundings. Now go out to a jobsite and look around. It's a little different perspective and necessitates the need of social cues. There are signs, backup bells, lights, hand signals, yelling, and radio communication. To get the job done, the workers must be in constant communication. You'll never see a jobsite foreman take a picture of a problem, walk all the way back to his trailer, and e-mail the superintendent, "Will you come look at this issue when you have a minute?" That's just not practical, so why is it considered practical when you're virtually building a structure? The reason is because BIM is technology driven and is challenged by the hypnotizing effect of the glowing screen. Modelers will default to e-mailing one another essays on an issue before they pick up the phone, which can waste time and effort, and can be misinterpreted. Technology has made many professionals believe it's more effective to send verbal communication in a digital form. However, not all e-mails are bad; it depends on the situation.

Different avenues of communication can be used effectively depending on the purpose of the communication and the clarity of information. Robert Lengel and Richard Daft write about these different avenues in their paper "The Selection of Communication Media as an Executive Skill" (*Academy of Management Executive*

[1988, 2(3)]). They say that "each channel of communication—be it written, telephone, face to face, or electronic—has characteristics that make it appropriate in some situation and not in others." This theory is commonly referred to as the *media richness theory*. For example, writing an e-mail to document an event when there's little room for misinterpretation would make sense.

> During our meeting on Wednesday 9/16/15, Innovative Architects agreed to export 2D CAD backgrounds every Tuesday.

This information is direct, documents a task, is easy to understand, and can be sent to all affected parties in one click. However, the majority of conversations that need to occur while virtually coordinating a building are more vague.

> If I move my duct 5" north in the south corridor, will that give you enough room to route your fire sprinkler main over the plumber's storm drain?

This information is convoluted. What part of the south corridor? What storm drain? It involves multiple parties (mechanical, fire protection, and plumbing) and, depending on the answer, could lead to e-mails back and forth between all parties involved. This type of communication is better handled in a social atmosphere where feedback can be delivered quickly between the parties. Figure 2.15 shows that as social cues deteriorate, the information's clarity must increase in order to maintain effective communication.

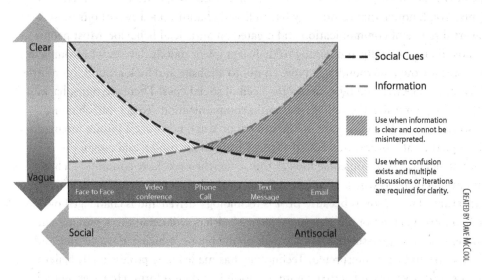

Figure 2.15 Effective communication

BIM involves communication and collaboration just like the jobsite. As shown in Figure 2.15, having the entire team co-locate in a space for coordination is the most effective way to virtually build a project, "because it has the capacity for direct experience, multiple information cues, immediate feedback, and personal focus"

(Lengel and Daft, "The Selection of Communication Media as an Executive Skill"). This is not always possible and, depending on the size of a project, may not be cost effective. BIM is a global technology, so the geographic location of team members can make face-to-face communication impractical. When working remotely, all team members should be able to easily access one another's contact information. This information should be located at the beginning of the execution plan. The Autodesk Communication Specification prioritizes the Core Collaboration Team on the first page of the plan after Project Information. Typically this contact sheet is a Microsoft Excel spreadsheet. Table 2.5 shows a contact sheet example.

▶ **Table 2.5** Sample of a People and Software Communication Sheet

Joe's Mechanical	Phone number	E-mail	Software and version	File exchange formats
Name: Jim Mynott	Office: 310-213-1234	jmynott@jmechanical.com	Revit 2015	RVT, DWG, IFC
Title: Lead detailer	Cell: 310-854-9654		Navisworks Manage 2015	NWC
Name: Brandon Kelly	Office: 710-456-8514	bkelley@jmechanical.com	Revit 2015	RVT, DWG, GBXML
Title: Detailing engineer	Cell: 949-219-8794		Navisworks Manage 2015	NWC

Based on a contact sheet example from McCarthy Building Companies, Inc.

Pick up the Phone

Phone calls are the most effective form of communication whenever a gray area exists in choosing the appropriate medium. For this reason, the direct number of every lead detailer/modeler *must* be identified on the contact sheet. Think through the following questions before sending an e-mail. If the answer is "No," then pick up the phone.

- Is the information clear?
- Does this need to be documented?
- Does this issue affect more than one person?
- Can I type this message out faster than I can talk it out?

Software

The other hurdle that the virtual building world faces is the communication between software systems. Let's take an example from a software program that most people are familiar with: Microsoft Word. You probably never noticed how many ways you can save a file in Word. Figure 2.16 shows the Save As drop-down options in a Word document.

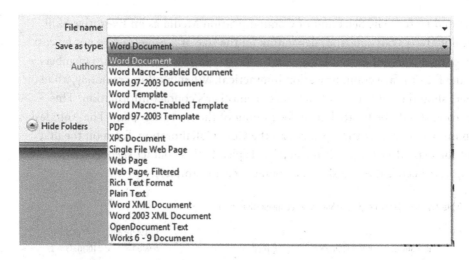

Figure 2.16 Save As options in Word

In Word 2010, you can save a file seventeen different ways. Each file format serves a different purpose. This same concept exists in BIM software. If you took Autodesk Revit, for example, you'd find that there are at least 13 different export (save as) formats besides the RVT, or native, format (see Figure 2.17).

Each format serves a different purpose in communicating with other BIM software platforms. Let's take GBXML (Green Building XML Schema), for example. This format is used to support model-based analysis for whole-building energy simulations. In our discussion of model-based analysis, I talked about the ebb and flow between the architect, mechanical engineer, and electrical engineer. In order for these three parties to know where to adjust their designs, they have to communicate information from a Revit model to an energy analysis program like Autodesk Green Building Studio. This is done by exporting their content from Revit (RVT) to the GBXML format. Once their content is imported into Green Building Studio, they can do an analysis to determine what adjustments they need to make to their design in the Revit model, and the process repeats until the owner's goal is met. This import-export communication barrier exists in all of the other fundamental uses of BIM as well:

- In model-based coordination, you might have to export as IFC.
- In model-based scheduling, you might have to export as DWFX.
- In model-based estimating, you might have to export as DWG.
- In model-based facilities management, you might have to export as the programming language ODBC (Open Database Connectivity).

Now that you're thoroughly confused, let's complicate it a little further. The various BIM platforms also have trouble communicating between versions of the software. This means that Revit 2013 will not read Revit 2014 files, so if the mechanical engineer was using Revit 2013 and the architect was using 2014, they would not be able to communicate back and forth unless they both moved to the more recent title. When you look at the combination of file formats needed to communicate on top of versions 2012–2015 and all the different team members on a project who need to communicate, you end up with the potential for a chaotic mess of versions and miscommunication, which will cause roadblocks. That's why the information exchange plan is an important part of the BIM execution plan.

Figure 2.17 Revit export formats

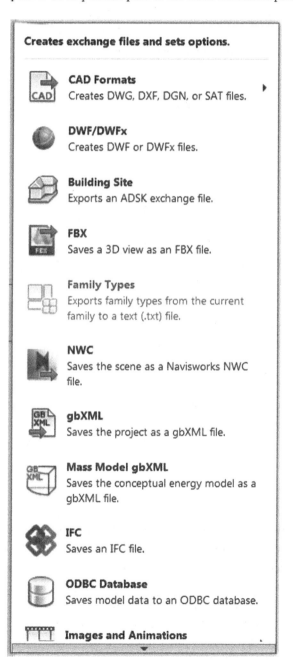

Creates exchange files and sets options.

CAD Formats
Creates DWG, DXF, DGN, or SAT files. ▶

DWF/DWFx
Creates DWF or DWFx files.

Building Site
Exports an ADSK exchange file.

FBX
Saves a 3D view as an FBX file.

Family Types
Exports family types from the current family to a text (.txt) file.

NWC
Saves the scene as a Navisworks NWC file.

gbXML
Saves the project as a gbXML file.

Mass Model gbXML
Saves the conceptual energy model as a gbXML file.

IFC
Saves an IFC file.

ODBC Database
Saves model data to an ODBC database.

Images and Animations

As a best practice, information exchange plans begin with the desired "end state" of information delivered and pull backward through construction and design to ensure that information is delivered at the right time and in the correct format. Because some people are visual-spatial learners and others are analytical-sequential learners, information exchange plans may be shown in different ways. Some people can understand the flow by reading, but some need to visualize the flow. Once the team members determine a required use of BIM, they must create an information exchange plan in narrative and/or visual form. The purpose of the plan is to describe where the information is created and how information will flow between team members (Figure 2.18).

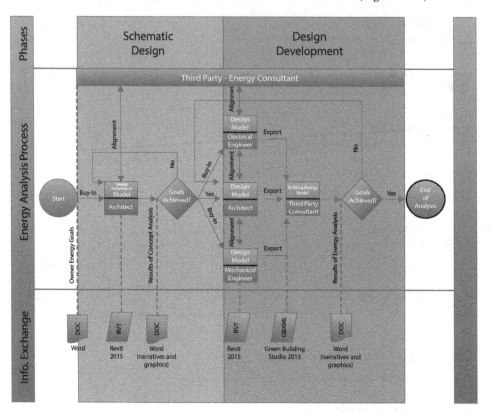

Figure 2.18 Information exchange plan

Information exchange plans don't have to be as complex as the one shown in Figure 2.18. Although this is a robust example, the point is that the team needs to fully understand the version of software and the file formats to be exported to ensure the different uses on the project will be successful. If not correctly planned for, what could end up happening is that the BIM manager will feel obligated to export file formats for everyone to keep the project moving, which hinders the manager's ability to focus on other tasks. Each team member needs to understand what is required to communicate across the various software platforms being used on a project.

In Table 2.5, earlier in this chapter, you'll notice that the contact sheet has personal information as well as software exchange information. The Autodesk Communication Specification separates this chart into Core Collaboration Team and Detailed Analysis Plan. However, it is useful to have them combined. It serves as a final check for communication and understanding. After the information exchange plan has been discussed and finalized (usually at the BIM kickoff), the BIM manager or information manager can send out a blank spreadsheet similar to Table 2.5 and have the lead modelers fill out and return it. The BIM manager or information manager can compile all the spreadsheets and quickly scan a single spreadsheet to make sure everyone is aligned on the version (2014, 2015, etc.) and the file formats (RVT, DGN, DWG, etc.) required for seamless communication between people and software. Once reviewed, the information exchange plan is inserted into the execution plan.

Expectation

At this point in your project-specific plan, you have identified goals; the personnel and software you need; what has to be modeled (the LOD); and an information workflow (information exchange plan). Randy Deutsch, in his book *BIM and Integrated Design: Strategies for Architectural Practice* (Wiley, 2011), states that's the easy part. "People-oriented factors are a greater challenge than solving the software, business, and technical problems of BIM implementation.... People are the crux—the key—to advancing BIM and integrated design." To have successful implementation, you have to address the human element. What are the expectations of the team?

Defining expectations is much harder than most people think and has a tendency to be overlooked or not communicated well on projects in general. In psychology there is a term called *selection bias*. Wikipedia defines it as "an error in choosing the individuals or groups to take part in a scientific study" (http://en.wikipedia.org/wiki/Selection_bias). The term roughly means that any given sample of individuals you choose has the potential to be unevenly balanced, which could skew the accuracy of the results of a scientific study.

One of the reasons why expectations are overlooked or miscommunicated in BIM is what I call *expectation bias*. It's defined as an error in expectations of the individuals or groups who take part in a BIM project. What tends to happen on a project is that due to the expectation bias, the BIM manager or information manager will unintentionally assume everyone has a basic understanding of what is expected for their project without ever communicating what the expectations are. This biased assumption could be due to a number of factors:

- Familiar team members
- Team members with robust resumes
- Team members with positive reputations among the BIM community
- An assumed understanding due to parties signing a contract that required BIM

However, you'll quickly realize that you're only as good as your weakest link and every job has a different approach. In a phone conversation with Craig Dubler, the project manager for Penn State's BIM Project Execution Planning Guide, he said: "BIM is only as good as the...least technical person. You bring them up or everybody else gets to that level." One person who doesn't follow or understand the procedures, shows up late to meetings, doesn't upload models on time, or doesn't maintain the coordination schedule will cause the whole process to stumble and affect the morale of the team. The expectations have to be universally understood and communicated. Do not assume that everyone understands your expectations. Communicate the expectations for every use of BIM and define them in the execution plan.

Each use of BIM requires different roles and responsibilities. Let's continue to use the energy analysis example. The information exchange plan may have shown that the third-party energy consultant oversees the process from SD to DD, but what does that mean? It hasn't been defined, has it? Is the consultant the right person to lead the effort? Maybe the consultant should just run the energy model analysis and the mechanical engineer should lead the effort. You'll want to discuss with your team past experience, lessons learned, and best practices. Remember, you're all in the boat and you're only as good as your weakest link, so it's in your best interest to get everything on the table up front.

How you organize the expectations of the team is up to you. It can be a chart, narrative, or a graphic, similar to the information exchange plan. The sidebar "Energy Analysis" shows an example.

Energy Analysis

BIM/Information Manager

- Organize weekly (Friday) meetings with Arch, ME, EE, and third-party consultant.
- Record meeting minutes and distribute them within 24 hours.
- Consolidate the design models using <*software*> for the team to review and visualize during the meeting.
- Create views in the model of issues discussed in the meeting. Upload the consolidated model to the FTP within 24 hours for the team's use.

Third-Party Energy Consultant

- Lead coordinator for energy analysis throughout the design process
- Direct team on how to model and what information is required in their model in order to get accurate results.
- Aid architect in running conceptual energy model using Revit.

- Attend all weekly meetings (Friday, 8:00 a.m.).
- Communicate with Arch, ME, EE, and information manager throughout the week.
- Assist the Arch, ME, and EE on exporting GBXML files as needed.
- Download files from FTP and run the energy model analysis every Thursday after 1:00 p.m. using <software> and print results for the Friday meeting.
- Direct team on changes required to achieve the owner energy goals.

Architect

- Run initial conceptual energy model during schematic design to optimize daylighting, orientation, and heat gain of the building.

Arch, ME, and EE

- Model systems per the LOD matrix.
- Export and upload GBXML files to the Energy Analysis folder on the FTP every Thursday before 12:59 p.m.
- Attend all weekly meetings with lead modeler.
- Communicate with third-party consultant and information manager throughout the week.
- Document the design changes made throughout the week to summarize them in the weekly meeting in order to analyze cost impacts.

Every use of BIM will require a description of roles and responsibilities, because the expectations will be different. For instance, the third-party energy consultant will not be involved in model-based scheduling, so these maps are not universal. Also, if you think back to Robynne Thaxton Parkinson's quote about "less is more," you may realize that some of these expectations are in the governing contract, BIM addendum, or information exchange plan, so these expectations may just say, "refer to _____ in the BIM addendum." That is a leaner approach and I find that it's a good way to make sure people are reading the other documents. The actual "engineer" is not always the person modeling or exporting the model, which is why it says "attend all weekly meetings with lead modeler." Many times the lead modeler isn't involved in the contract negotiations, so by putting "refer to _____" it will flush out if the engineer has shared the addendum with the lead modeler. Ultimately, you want to make sure everyone understands the plan and expectations, because you will have expectation bias on your project, guaranteed.

Organization

The last piece of the execution plan is *organization*. Organizing BIM is similar to organizing a toolbelt. Every tool has a place so you can work quickly and efficiently.

Experienced craftspeople often get to a point where they no longer look at their belt to get what they need. The desired goal for organizing the model and files being used on a project should be intuitive. The team shouldn't have to waste time searching for the tools (files) they need.

Model Origin

The organization of BIM starts at the *origin*. The E202, E203, and ConsensusDocs reference what the roles of the BIM information manager are. The top of the list in all documents is that the BIM/information manager defines the origin of the model. If BIM is going to be used collaboratively, the models must align. The origin is project specific, so it shouldn't be in the contract or the addendum—it has to be in the execution plan. The coordinates in BIM work on axes, so X = North/South, Y = East/West, and Z = Elevation, so defining this in the execution plan is as simple as stating, "The model origin for this project is X,Y,Z." However, some teams prefer world coordinates based on state planes, and some teams prefer defining an arbitrary 0,0,0 to the model. This should be discussed in the initial goals, uses, and objectives discussion, because if you're planning on leveraging the model throughout design, construction, and maintenance, the model has to have relation to world coordinates.

Model Storage

The second role of the BIM information manager, based on the AIA documents and ConsensusDocs, is to organize and store the files on a BIM project. A number of file management software programs, FTPs, and cloud-based solutions are available. As with most software options, choosing one can be a bit overwhelming. Although some integrated firms may have the luxury of consistency, most owners have unique file management programs that may require you to use different systems. The key here, as with any BIM decision, is to develop a solid process and let technology enhance it. Don't let technology drive your process, because it will constantly change.

Regardless of whether you're using your own model storage or the owner's, be sure to define the solution and explain how the team gets access to the files in the execution plan. The majority of the time, team members are invited to the FTP or management software and then they're required to do some setup on their end to get access. This setup process, whether it's creating a username and password, should be spelled out in the organization section of the execution plan.

Folder Structure

The folder structure needs to be based on the uses. For instance, what file would the third-party consultant for energy analysis be most interested in? Based on the information exchange plan and the expectation narrative, the consultant is only interested in the GBXML files, so it makes sense to create a specific folder,

Energy Analysis, under the Model-Based Analysis folder. This way, the Arch, ME, and EE can upload the GBXML files to that folder so that the consultant knows exactly where to find all the files he or she needs. The folder structure might look like this:

- **01** Model-Based Coordination
- **02** Model-Based Scheduling
- **03** Model-Based Estimating
- **04** Model-Based Facilities Management
- **05** Model-Based Analysis
    - **A** Daylighting
    - **B** CFD
    - **C** Energy
        BLD-ARCH-ALL.gbXML
        BLD-MECH- ALL.gbXML
        BLD-ELEC- ALL.gbXML

Avoid too many individual Arch, ME, and EE folders or root "dump" folders to store the RVT, DWG, IFC, and other files because searching different places and sorting through the various file types could prove time-consuming to a team member. Not only is this inefficient, but it opens up security concerns. Most storage solutions allow you to assign different levels of accessibility rights to the users (download permission, download/upload permission, download/upload/delete permission). Third-party consultants shouldn't have access to modify or delete the Revit models. They should only have access to modify or delete content they are responsible for.

Every use should be analyzed the exact same way so your toolbelt can be organized efficiently. Don't mix your screws, nuts, and nails. Once you have organized the folders for the job, take a screenshot and insert it into your execution plan so that everyone is familiar with the folder structure for the project. You may also want to briefly describe the content that should be stored in each folder.

Note: Once you have developed an organized toolbelt (folder structure), save it as a template for use on another job so you don't have to start from scratch every time.

File Naming

The last part of organization is naming the files that go in the folders. Don't overthink this. The filenaming should be simple and identify three main things: the project, the author, and the zone. Most storage solutions have the ability to automatically version or archive files. This means that you can upload a file with the same name and it won't overwrite the previous one—it will just create a newer version and archive the old. This has made the BIM process much more efficient. Models are constantly changing throughout the design, so it's important for the team to have access to the latest files. If files have inconsistent names or people start adding dates to the end of the names, the folders become a cluttered mess of files and someone, without fail, will be working off an old file. If you have a consistent name and the storage solution automatically versions, your folder structure will be simplified and the risk of loss eliminated. See the sidebar "Filenaming Convention" for an example. It discusses a basic example of a naming convention that covers the essentials of what should be in the filename. Filenames could get much more complex, but be careful about how many characters you use. You want to make them simple to understand, so having a name like 76543.087-Hospital-BLD-E-EL-02-Z1 (Project Number 76543 .087 - Hospital Building "E" - electrical content - Level 02 - Zone 01) would be a bit of overkill. Keep it simple.

Filenaming Convention

Project Abbreviation: BLD1

Author Abbreviations

- AR = architect
- ST = Structural engineer
- ME = Mechanical engineer
- EE = Electrical engineer
- PE = Plumbing engineer
- FP = Fire protection

Zone Abbreviations

- 00 = Underground content
- 01 = Level 01 content
- 02 = Level 02 content
- 03 = Level 03 content
- RF = Roof content
- ALL = All levels

BLD1-AR-02 = Architect's content for Level 02 in building number 1

Similar to the folder structure, once a good naming convention is created it can be saved for use on future projects. Having a consistent naming convention for all your jobs will allow for efficient flow of information, communication, and documentation.

Summary

This chapter demonstrated the amount of effort and analysis it takes to plan for BIM on a project. The general statement "We're going to use BIM on this project" should carry much more weight. The answers to the following questions will guide you in developing a plan for success:

What's the delivery method?

Where is the AE team in the design process?

Is there a BIM contract requirement?

Are the designers modeling?

What's the level of development?

Are they going to share the models?

Once the draft of your plan is developed, all team members will need to review it in order to have a "BIM kickoff." This meeting is where you discuss the goals, uses, and plan for executing the project. This meeting should encourage collaboration and allow for open discussion of best practices. Technology will continue to advance faster than any one person can keep up with, so working as a team is the best way to discover new efficiencies.

Thinking through and drafting an execution plan will also prepare you for interviews before being selected for a project. More and more owners are requiring BIM personnel to be in attendance at the interviews and are allotting a significant amount of time for companies to demonstrate their BIM capabilities. So, if your BIM managers haven't been to toast masters, then it may be time to sign them up for some classes. The next chapter explores how to market BIM and win projects.

How to Market BIM and Win the Project

3

How do you win work through the innovative use of BIM? How do you select the right project partners? How can you use technology as a differentiator for your team? The answers to these questions are a pivotal part of building a winning team and being selected for today's construction projects.

Construction procurement criteria have moved beyond price-only selection models. Smart clients, frustrated with the results of previous contract award methods, have found new ways of selecting a construction partner. These clients are keenly aware of project complexities and are looking for firms that use technology to overcome these challenges.

In this chapter:
Bim Marketing Background
Building Your Team
Marketing Your Brand of BIM
Using Bim to Enhance the Proposal
Client Alignment
Seeking Value and Focusing on Results

BIM Marketing Background

Why would I dedicate an entire chapter on marketing and selling BIM?

For starters, there are new factors to consider when having discussions with potential customers and project partners that can create elements of focus, trust, and extended value. In the early 2000s, if your construction management firm was using BIM, you probably would've had a tool that was viewed as innovative among competitors. The use of this new technology and the hope of delivering construction projects in a more effective way weighted many decisions favorably to these teams who were tech savvy and forward thinking.

Although technology and BIM increased the salability in construction, it also had an underlying effect. The introduction of BIM began to shift the dialogue in a once stodgy industry from "This is the way we've always done it," to exploring innovation in other areas and asking "Why not?" Generally, owners who selected construction management firms with new technologies discovered new ways of working together. These new and more effective methods of collaboration were generally ignored, and BIM was jammed within the same processes as before. This wasn't entirely negative. Many owners saw the potential opportunities with this promising new tool, but saw its effectiveness diluted because of how it was being used within traditional processes. As a result, many owners shifted their preferred delivery methods and began to ask for metrics and case studies of successes and lessons learned as a measuring stick of a team's effectiveness with technology and processes in a results-driven manner. In many cases, BIM is responsible for the technological renaissance the construction market is now experiencing. Tandem to BIM's growth in industry adoption, the momentum of technology innovation in construction is widespread, with the focus now shifting to applications that create value and measurable results.

The ability to use BIM during construction has been an ever evolving process, and it continues to adapt and change through the construction process. As mentioned in Chapter 2, many proposals and RFPs in the early 2000s included requests from owners prescribing "do BIM" or "Contractor must use BIM during construction." These requests reflected a lack of mature understanding of the potential value for BIM from owners. Additionally, construction management firms had no clear understanding of its uses and how it should be integrated into a project. At this stage in BIM's evolution, there was an industry shift. Some owners and construction consumers were deeply ingrained in BIM and were pushing the envelope early into the BIM adoption cycle. These early owner-adopters were a major contributing factor to the adoption of BIM throughout both the design and construction management landscape because of their position as customer and power to steer industry focus. However, the lack of standards and guidelines for owners at that time presented a series of needs relating to information and syntax consistency as well as clarity in the process of using BIM effectively. Because of the developing nature of this technology, the way BIM was requested or contractually required as a deliverable varied among owners—each had their own perspective as to what would be valuable for them throughout design and construction.

As BIM began to become a major industry trend (somewhere around 2007) and the early majority of owners began to better understand BIM uses from their peers, a common language emerged about BIM and the way BIM could be used throughout the design and construction process that would create more value than a traditional CAD process. Just as important as portfolios, staff quality, and safety, teams began seeing language in RFPs and contracts that require information management plans, BIM plans, and definitions of how work is to be managed in the field. Winning teams found ways to integrate the use of BIM and other technologies with traditional construction procurement factors to strike a chord with customers, align better with their partners, and win more work.

In the same way owners were requesting enhanced use of BIM on their projects, contractors were learning how to adapt and compete in this new BIM-enabled frontier. Unfortunately, many owners were exposed to overzealous commitments made by construction management firms during this stage that overpromised and underdelivered in order to get work. Contractors claimed that they had "fully integrated" processes and software, when in reality they were not fully prepared or experienced enough to deliver the project effectively and thus fell short of meeting customers' expectations. This phenomenon of overselling or "BIM-washing" (Figure 3.1) created an industrywide dialogue to address this issue. Thankfully due to many associations and organizations, owners were able to discuss among themselves and compare the quality of deliverables they were getting from various firms, which promoted the companies who were using these new transformative tools and processes over their competitors who were simply paying BIM lip service.

Figure 3.1 "BIM-washing"

Once the industry realized that BIM was not just a fad and was becoming a core part of a new way of doing business, it saw an honest dialogue of best practice sharing, lessons learned, innovation, and an exciting paradigm shift that has transformed the way construction is delivered. Moreover, as BIM became the new norm, industry associations, organizations, and peer groups began focusing some of their efforts to continue the dialogue of BIM and technology in the construction management industry. To say BIM is now "table stakes" in its adoption globally is probably an overgeneralization of the actual condition. The reality is that while many construction management firms have now been using these tools and processes for some time now and are bringing to bear an enhanced value proposition for construction consumers that extends beyond the physical structure to data structure and information management, others are just beginning to use some of the tools.

"Doing BIM"

I'm always leery when I hear someone say, "I do BIM" as an all-encompassing statement. Jokingly, I equate this statement to saying, "I do the Internet." Having dedicated the majority of my career to BIM and technology in construction, I believe we are just scratching the surface with what the capabilities of BIM and technology hold in our profession. However, I don't believe the new innovations will be model-centric—I believe they will be information-centric.

In fact, as users become increasingly proficient at modeling, there will be a point (just as in CAD) where user expertise peaks (a model can be created only so fast) and BIM will be accepted as the way to do design and construction work. Additionally, users will explore use cases outside of these traditional deliverables and look for extended value in the information.

This value is being leveraged to increase fees and productivity for many design teams as well as to mitigate risk and increase margins for construction firms by streamlining items such as procurement, quality control, inventory, prefabrication, and many other applications. So though a user may be using some aspect of a BIM tool or process, I think our industry still has a long and exciting path in front of it to find the deeper potential for information within these models.

Building Your Team

In the 2014 movie *Draft Day*, Kevin Costner plays Sonny Weaver Jr., the general manager for the Cleveland Browns. The plot of the movie is based on the trades and positioning completed by Sonny around the NFL draft. During the movie, Sonny thoroughly analyzes the past performance and attitudes of an array of players to ultimately get the draft picks that he believe will perform best for the Browns. In the movie, the character everyone believes will be selected by Sonny as a first draft pick, Bo Callahan, seems to be the perfect player. Yet as Sonny digs deeper he realizes that

Bo isn't necessarily the team player that everyone thought he was, and that his attitude puts personal gain ahead of the team's success. Ultimately, Sonny uses his first-round draft pick to select a different player with a deeper team focus and a better attitude to the shock of the world and the rest of the NFL teams who had anticipated otherwise. The parallels between selecting the right team members, both as an organization and when partnering with external companies, hold true.

Attitudes are everything and can make the difference between having professionals with the mental endurance to tackle a difficult issue, move out of comfort zones to learn something new, and improve versus having employees who just show up for a paycheck. Today's construction teams are learning that in order to win, enabling behaviors are critical to achieving success.

The landscape of design and construction has changed from that of 30 years ago. Instead of considering construction solely as a commodity and price-driven market, owners, design firms and construction companies are shifting to more collaborative ways of working together and completing projects through value-based selection and shared incentives. There are still examples of price-driven construction projects, such as developer-driven first cost jobs, municipal work and some industrial and manufacturing projects. However, the continued push of cost as the main criterion for project selection works against a project team in many ways. Understanding that human behaviors play a critical role in construction, and as mentioned in Chapter 2, integrated methods of delivery will continue to gain prominence. So how does hard bid work against a team—especially when a team selected on price may not have any prior experience working together and may not be familiar with one another's systems or possess the capabilities and tools required to collaborate with one another?

Price-based selection team members are not necessarily rewarded for being a collaborative partner in the design and construction process and keeping their costs low. Think about a subcontractor who bids on a project with razor-thin margins— do you think she will willingly accept small changes and not charge more for any additional scope? Of course not! Her firm was selected based on price, and as soon as anything is added, that price goes up. Now, think of an integrated Design Build or Engineer, Procure and Construct (EPC) team, which has joined together and mutually submitted a not-to-exceed budget to a customer. Is there a benefit to them figuring out issues internally, without going to the owner with their hands out? Absolutely! Because these firms realize that going to the customer for more money is typically not an option in an integrated, value driven project. Mutual accountability aligns teams and drives at project-focused solutions rather than individual firm benefit, because shared value and contractual alignment focuses on achieving a better project as a team. So how does technology fit into integrated teams, whether selecting the right internal staff or deciding who to partner with outside of an organization? Simply put, teams must determine how they want to work first, select partners that can meet those demands, and develop a cohesive technology strategy that meets the project's requirements.

When building a winning technology team, it is important to initially consider multiple factors. As a general guideline, consider these issues when selecting team members internally or externally:

Delivery Method   Which team members have proven to be good partners in this delivery method? Are there innovative offerings within this method that allow you to be more competitive?

Project Schedule   What do you have time to achieve using technology and which team members add value or have experience in using BIM or technology within that timeframe?

Technical Expertise   How advanced do users need to be to meet the demands of this project? Is the project highly complex and will it require certain skill sets?

Client's Technology Requirements   What is the client asking for and who can best deliver that? Do you have the expertise or do you need to leverage someone else's experience?

Why go through all this trouble for a marketing or sales effort? Because construction management is a service-based offering and companies "sell" their staff's capabilities and experiences. Just as an architecture or engineering firm markets their firm's brain trust to a customer, construction companies market their personnel to customers. For this reason, among others, it is important that the personnel proposed for future work fit the projects' requirements and create the ability for a firm to differentiate themselves among their competitors.

Increasingly, personnel such as BIM managers, technologists, and IT professionals are participating in project pursuit efforts to address unique client and project needs. BIM specialists may share enhanced visualizations to display a better understanding of the project, technologists may propose custom applications that make the client's job easier, and IT staff may describe how remote networks will be set up and how connectivity will be established to ensure production on-site. This type of alignment works well with customers and in extremely close project pursuit efforts, may be the deciding factor for the job. Later in this chapter, I'll discuss in detail how to look at various approaches of marketing to a client with BIM and technology.

Keys to "Talking BIM"

"By using BIM-derived RVT data, he extracted an XML schema that was then used to populate a WBS DB and inform our AWPS via an API for our 5D tool."

We all understood that, right?

Although your colleagues may understand what you're saying, your customer or audience may not. Here are some tips for speaking better BIM:

Death by "BIM-cronyms" Limit the acronyms use the full term and don't assume understanding on the other side of the table.

Know Your Audience Are they tech-savvy or not so much? Speak to the least informed member of your audience.

Simplicity Shows Mastery People appreciate elegant concepts and clearly stated workflows. If you don't understand something, don't pretend you do. It's okay to say "I don't know"—it allows you to learn something new.

Talk to Processes when Possible Everyone appreciates stories to help illustrate concepts and toolsets. Use them with reckless abandon.

Marketing Your Brand of BIM

As I've said before, BIM for construction in simple terms is *virtually constructing a structure prior to physically constructing it.* This effort includes simulations, analysis, dissection, model refinement, and collaboration. Within this effort is the opportunity to define and refine these processes in depth, the resulting outcomes, lessons learned, or discovered innovations. These input points shape the use of BIM within a firm, and for some, these refined workflows become a differentiator.

When a construction management firm is asked the question, "What is the best way to market how we work?" there is usually some degree of pride and achievement behind the answer. Many firms have developed their own processes of using tools that make them unique. The best teams in the market are no longer rattling off the list of tools that they have loaded up on to high-powered laptops as an answer for how they use BIM. Instead the discussion is shifting to understanding and aligning to desired customer deliverables, and what processes a team must go through to have these discussions with owners early in the process to optimize the value of BIM throughout design and construction. Additionally, firms are creating better ways of working together. In many cases, this involves the use of cloud environments and real-time collaboration tools that close a feedback loop and pull information from a system faster to make better decisions.

Regardless of what technology solutions a firm decides to use for a particular project, there needs to be a focus on customer value that asks, "Are we selecting the *right* technologies for this client rather than showing *all* the technologies we have?" Though it's meaningful to educate and share with owners what certain technologies do, it is also important when you are in a pursuit strategy to show expertise and value in the tools and workflows you propose. Just as each construction project is custom, so should be the choice of tools.

In short, a best practice for marketing your brand of BIM should revolve around five key factors:

- Make sure what you are proposing shows clear and demonstrable value.
- Clearly state whether this is a proven tool or process, a developing one, or an innovative one.
- Show real results from the impact of implementation.
- Give the owner what he or she wants or clearly state why you can't.
- Make sure you are offering something that you can deliver.

Though some of these factors might come across as common sense, there is a consistent thread of logic that is tied to each one of these questions in defining whether a particular technology makes the cut to be included in an RFP response or a proposal to an owner. The exercise of thinking through these questions makes for more meaningful content in a proposal, as well as showing how an owner's needs are understood and addressed.

Does What You Are Proposing Show Clear and Demonstrable Value?

This question gets to the core of what BIM strives for. Another way of asking this question might be, "Is what you're doing creating a better project?" Answering the question of whether a tool shows clear value sounds simple enough. However, many construction management firms make the mistake of forgetting their audience, and though a team may be an expert at a particular tool or process, it doesn't necessarily mean that an owner is equally educated or understands the value proposition. The use of acronyms and complex systems doesn't help either. The information doesn't need to be "dumbed down," but it should be clearly illustrated as to what issue(s) a proposed tool and/or workflow fixes and why the outcomes are better than another method. As economist and author E. F. Schumacher says, "Any intelligent fool can make things bigger, more complex, and more violent. It takes a touch of genius—and a lot of courage—to move in the opposite direction."

What is proposed shouldn't just be tools, but rather the way a team thinks through delivering a project that can make it stand out. For example, the execution of a well-written BIM execution or information transfer strategy might show a team's expertise in using both the tools, as well as uncovering a deeper understanding of the customer's needs and how the team proposes to work together. This plan may give the team a competitive edge over another team who shows an array of tools with no clearly developed strategy into how the information will move, who has responsibility, or how work will be executed.

Another consideration in asking the real value question is to think through how information will "flow" through the project and how information created in one place can either be connected to other systems or at least leveraged to reduce manual input times and redundant efforts. As the team is coming up with answers to the question of clear value, consistently ask, "*So what?*" Does what you are proposing really matter to the customer? Will they see value in it? Is it important to them? The answers to these questions will help steer you and your team to a customer-centric presentation and one that seems custom built for them—because it was.

Is This a Proven Tool or Process, a Developing One, or an Innovative One?

Why is it important to identify the stage of adoption for technology tools? Because tools introduced in the wrong context may create confusion and risk to a project team. Many times a firm or team makes a call to introduce a particular "widget" that looks "cool," attempting to set apart the firm from competitors. Yet the tool provides no clear value to the process or project overall. In many cases, the introduction of new technology into a project without thorough consideration of how it will help creates risk for the team. Keep in mind that once a tool is shown and the owner is exposed to it, there is often an expectation of use. The potential risk of "BIM-washing" or "Hollywood BIM" can do more harm than good in a project.

This is not to say that a firm shouldn't introduce an innovative tool during the selection process, particularly if it could create significant value for a customer. However, make sure the correct context is used when describing the tool and be straightforward in your amount of experience with the tool. Many times, owners like the idea of using a project as an opportunity to innovate or test new solutions because if they work, the owner can require that technology on future projects. However, owners need to know where a tool is in its adoption cycle to know what to expect.

Proven tools and processes are the foundation for development in a firm using BIM. For example, an owner may want to better understand how a firm will mitigate risk in trade coordination or inventory management. Having the ability to walk an owner through proven processes and show the outcomes is invaluable and establishes the credibility of a firm's ability to perform. Additionally, proven outcomes also allow a firm to give examples of previous projects where workflows created value, mitigated risk, or saved the day. In marketing BIM and technology, the more a firm can show with examples, the better. Yet, it is also important to show how these tools connect and talk with one another. As firms become proficient at a particular deliverable, it is always beneficial to show how they continue to push the envelope and look for improved ways to deliver work product.

Developing tools and processes are often the hidden gem where there has been some degree of use or piloting though a tool or process may not be fully proven yet.

Developing workflows offer the ability to show an owner a construction management firm's progress in determining the value of new tools and processes. Quite often in RFP responses, and even in interviews, the ability to show the rigor and study of testing various tools provides insight into a firm's innovation cycle and defines what a firm believes will be valuable for its business and its customers. This clarity can create a common alignment with owners; in all likelihood they have experienced similar strategies in making the efforts needed to stay up on the latest technology to create value.

Selecting innovative tools and developing new processes are the fun parts of using technology in construction. They allow the opportunity to shift mind-sets from traditional methodologies to look at better ways of working together and building systems that streamline the flow of information. In some respects, innovation is a tricky business because there are so many potential ways of doing the same thing. Usually it's a combination of tools, processes, and behaviors that can be adjusted to alter results. Most companies look to innovation as a major component to sustained growth and value.

Where innovation occurs is a much larger subject than I'll cover in this section, since it pertains to a host of factors and motivations, some personal, some business driven, and some out of curiosity. The AEC industry at large is looking at ways to improve the visibility and value around innovation in events such as the AEC Hackathon (Figure 3.2) and the ENR FutureTech Conference that bring together bright minds from all areas of the design and construction industry to investigate and solve problems through the use of technology. Events such as these remove competitive barriers between firms and allow for cross-functional teams to apply their experience and knowledge to solve common issues that persist in our industry.

Figure 3.2 AEC Hackathon

Innovative ideas come in all shapes and sizes. Some changes are smaller in nature and may represent incremental improvements. In lean terminology, these iterative changes use Kaizen or constantly improving systems with the goal of perfection. These incremental changes may appear to be insignificant when looked at individually, but when viewed as a series of iterative improvements or, even better, as a culture of innovation that rewards these ideas, there is potential for marketing the progressive nature of a firm.

Other innovations are big shifts that suggest that some traditional construct needs to be challenged to provide superior value. These large-scale ideas may challenge something fundamental, like the way an estimate is generated and issued or how schedule information is received from the field. In large-scale form, these innovative ideas have a history of changing the game in the way construction is delivered. Many owners have rallied behind new ideas and concepts offered from a project team and look to the better use of BIM as one of these shifts for mitigating risk on a project (Figure 3.3). Again, it's important to know the customer and ensure that a proposed offering is appropriate. However, in many cases the use of innovation in a project can set a firm apart from the crowd and get a team excited about the opportunity to use their project as a platform for industry growth.

Impact of Strategies on Mitigating the Seven Top Causes of Project Uncertainty
(According to Owners, Architects and Contractors)

Source: McGraw Hill Construction, 2014

■ Scores Above 80   ■ Scores 70 to 79   ■ Scores 60 to 69   ■ Scores 50 to 59   ■ Scores 40 to 49   ■ Scores Below 40

	Owner-Driven Changes	Accelerated Schedule	Design Errors	Design Omissions	Construction Coordination Issues	Contractor-Caused Delays	Unforeseen Conditions	AVERAGE
Better Communication Among All Project Team Members in Early Stages of the Project	88	96	94	88	93	79	79	88
Greater Leadership or Involvement by Owner in All Stages of Design and Construction	81	83	78	71	59	53	73	71
Use of Team-Based Alternative to Design-Bid-Build	64	70	75	71	72	49	65	67
Appropriate Contingency Dedicated to This Issue by Owner	79	70	73	48	54	57	79	66
Use of BIM	53	64	76	69	76	47	55	63
Shared Liability Across the Project Team for Problems Created by This Factor	48	59	71	62	63	53	58	59
Use of Lean Design and Construction Practices	28	48	32	31	39	32	28	34

IMAGE © McGRAW HILL SMARTMARKET REPORT

Figure 3.3 Impact of risk-reducing strategies

Can You Show Real Results from the Impact of Implementation?

One of the single largest challenges with BIM is showing hard metrics that justify the return on investment (ROI) for its use. Capturing metrics against a baseline would be much easier if you were using BIM in a manufacturing setting or a more controlled and redundant environment. However, as each construction project is unique, it's often difficult to justify savings over typical methods because it can be argued that traditional methods may have worked just as well.

In many instances of BIM, ROI can be explained in terms of time and money saved. These metrics may include faster turnaround times to complete analysis, increased efficiency to send and access information, and reduced hours dedicated to particular tasks. Although these metrics are solid and quantifiable, the more esoteric ROI questions that contribute to a project's success, such as increased collaboration, better visualization, and better quality, are more difficult to address. Project team members may validate some of these assumptions through their experiences in working together on projects through case studies and surveys. Aside from these "fringe" metrics, BIM continues to be accepted, even if some benefits are not measured. So how can the use and resulting impact of BIM be measured on a project? One answer is in the deliverables.

The easiest way to determine the impact and ROI of implementing a technology is to compare the time and money that went into producing a new deliverable versus a traditional deliverable. For example, say your team is considering integrating a new mobile application technology that would allow field staff to generate and respond to RFIs, punch lists, and material tracking through a tablet PC. By looking at a traditional method of this workflow, you can use a matrix chart to see that comparison in a simple way. You can use a matrix like the one in Table 3.1 to show the proposed value of a tool when pursuing work and proposing new technologies as well as to validate whether the assumptions made during the pursuit phase were accurate and whether efficiencies and benefits are being realized downstream.

Analysis at this level sets clear expectations for how a tool should be used and what should be achieved. These expectations can then be checked as the project progresses. In any construction project, some activities can be improved upon, which can then be evaluated to see if they met expectations.

The analysis aspect of answering the question of "Can we show real results from this implementation?" is one part of a two-part equation. The other major component is using experience and case studies. Examples of previous successes and integration and related stories of outcomes help validate efforts and best answer the question when marketing BIM or other technologies. Of course, the best solution is to substantiate these claims with outside input such as client or subcontracting partner feedback that can further reinforce the outcome of a decision or process. This input is not only powerful in marketing, but it is also a powerful tool in learning how to improve.

▶ **Table 3.1** RFI technology comparison matrix

Description	Traditional	Proposed (app)	Benefit
RFI creation	Photos are captured in the field on a camera and uploaded to a laptop. Drawing markups are made on a copy and scanned as a PDF. Specification clarifications are highlighted and attached to the description of the RFI. Approximately 10 minutes per RFI.	RFIs are created from the mobile application. The app attaches pictures directly from the device's camera roll. The app links to drawings and specs stored on a cloud server and can be accessed directly from the tool. Approximately 4 minutes per RFI.	The mobile app will save time in uploading and attaching photos and referring to documents. However, Internet connectivity is required and users will need to familiarize themselves with the new tool. Saves approximately 6 minutes per RFI.
Punch lists	The CM* walks the project and identifies issues by marking issues on a printed plan. Additional photos are taken with a camera and associated with the resulting Excel spreadsheet issued to subcontractors. Approximately 9 minutes per punch list item.	The CM uses the app to open a model of the project where issues can be tagged in 3D. Photos are taken from the application and hosted directly with the associated punch list item. Additionally, the tool automatically logs and numbers the responses as well as tracks response times. Approximately 4 minutes per punch list item.	The team will pick up efficiency both in the creation of the punch list and in response times by providing a single interface that team members can use to see and respond to issues. The tracking ability will ensure that reasonable turnaround times are being met. Saves approximately 5 minutes per punch list item.
Material tracking	The CM meets with the project's subcontractors to identify materials needed and scheduled deliveries in Excel. As materials arrive, their location and quantity is verified and individually tracked with a barcode. The barcode software requires the use of a special scanner to manage inventory. Approximately 15 minutes.	The CM uses the app to input the materials and delivery times. The CM receives a reminder when materials are to arrive. When they arrive, inexpensive barcodes can be attached to materials and scanned using the tablet's camera feature. Photos of damaged material can be associated with materials as well. Approximately 7 minutes.	By using the app, the team can reduce the time taken for logging and processing materials. Additionally, the application may save the team money by not having to purchase scanners and inventory software. Savings of 8 minutes per material package tracked.

* CM = construction manager

Is This What the Owner Wants?

Don't fall prey to the temptation to apply a cookie-cutter solution for every customer. This approach is lazy and you miss the opportunity to take what makes a particular project unique and capitalize on it through the use of a focused technology strategy. This is not to say that a firm shouldn't standardize their platforms, but in selecting which tools and processes will be used to execute a deliverable it is important to consider the question, "Is this *really* what the owner wants?" In many cases, it may be a more efficient way of working that benefits the construction management team the most, and this is fine. However, in pursuing work it is the teams who take into account the owner's needs and deliver customer-centric product that will make one firm stand out.

Firsthand Owner's Experience and Vision of BIM

The following was written by David Umstot, cofounder and president of Umstot Project and Facilities Solutions in San Diego, California:

"From the perspective of a 10-year public agency owner and 20-year practitioner, building information modeling has the power to enhance collaboration of the project delivery team, reduce coordination conflicts, optimize building performance, improve schedule performance, enhance the efficiency of building maintenance and operations, and reduce total cost of ownership. BIM may be the most powerful change in our industry in a century. By its very nature, it is not only a model loaded with information, but a system that fosters collaboration among architects, engineers, specialty trade contractors, and builders.

"We used BIM to help launch lean project delivery at the San Diego Community College District while I was vice chancellor in 2008. We recognized the potential of BIM to be a game-changing process and were early adopters, requiring it on all projects since 2008.

"Construction in the United States has the dubious distinction according to U.S. Department of Commerce data as being the sole industry to decline in productivity since 1964. BIM, in concert with lean project delivery, has the potential to greatly enhance the productivity of our industry. Metrics from our program based on 35 completed projects totaling $584 million in construction value indicate overall change order rates dropped from 7.73 percent to 4.43 percent of original contract value using BIM and lean methods, as opposed to our previous project delivery approaches. Design and coordination errors and omissions-related change orders similarly dropped from 2.99 percent to 1.88 percent.

"In the United Kingdom, the British government has set construction 2025 goals to reduce project costs by 33 percent, reduce overall project schedule from inception through completion by 50 percent, and reduce greenhouse gas emissions by 50 percent. BIM, used in concert with lean enterprise and project delivery, is the only way these monumental goals can be met.

"I also look forward to the transcendence of BIM into the realm of facilities management to enhance the ability to plan and deliver maintenance and operations in a more efficient manner and reduce the total cost of ownership. We are in a time of transformation in our industry and BIM is, and will continue to be, a significant contributor in this change."

An example of a customer-centric offering is to take into account the role of owners on a project. How do their workflows and processes align to what is being proposed? Is a tool that is being considered complicate or simplify their work? Is information flow being achieved, or do the systems offered take a considerable amount of rework or redundancy? Asking these questions will help to add a solid foundation of value to a proposal, as well as better prepare the team to address questions that may come about from the customer who may want to select a team who has considered their systems and workflows and has strategized how to dovetail their proposed solutions into the way they work.

In the current market, many owners are looking beyond projects being delivered on time and on budget. In many cases, these parameters are simply baseline expectations to even be considered for the work. As a result, there is a trend and paradigm shift to added value and looking beyond a facility to address other owner issues that may be solved during design and construction. For example, a municipal customer building a bridge may have concerns about educating the public about bridge closings and alternate routes. This may mean a visualization opportunity for a construction team to help provide renderings or videos showing and educating the public on the project and being proactive in their approach to lane closures (Figure 3.4 and Figure 3.5). A private healthcare customer may be concerned about the impact of a job trailer on a tightly crowded site for a remodel. In this case, the use of tablet technology and real-time cloud-based collaboration may help. The customer advocate concept serves construction management firms well in determining other ways to help owners achieve the end product, but also makes the experience an enjoyable and engaging one through the use of well-thought-out tools and processes.

Is This Something You Can Deliver?

The worst thing a project team can do when pursuing a project is overpromise and underdeliver. Not only does this leave a bad taste in the owner's mouth, it also tarnishes the reputation of the team. A solid reputation takes years to earn but can be destroyed with a single bad decision or project. Because of the sensitive nature of reputations, it is important for a project team to ensure that the tools proposed after making it through the "value filter" can be delivered on. As a best practice, a team should be encouraged to investigate the tool as well as communicate with others who have used it to get feedback and to see if it is viable prior to proposal. The act of checking into the background on a new tool serves another purpose because it better informs the way the tool is used on a project as well as captures best practices from industry peers and colleagues.

Figure 3.4 Construction site simulation video rendering

Figure 3.5 QR code link to video of simulation

Sometimes after investigation and research the answer to "Is this something you can deliver?" is simply no. In this case, the answer should not be viewed as a failure. Too many teams look at the inability to deliver on a big idea as a mistake to be remedied. In reality, this is often the best thing that could have happened to a project team. It is far better to discover that a new tool won't be viable during the proposal stage than during a project where the stakes are significantly higher and promises have already been made.

There are cases when a tool is analyzed for value, investigated for viability, and included in a proposal and it still falls short. In these cases, it is best to address the problem openly rather than make excuses and work with project team members to find another solution. Especially when introducing new technology, it is important to put the introduction of this new technology in the proper context (refer to the section

"Is This a Proven Tool or Process, a Developing One, or an Innovative One?" earlier in this chapter) when talking with a customer. Say, "We think this new *<insert technology here>* will create efficiencies and significant value for this project because *<insert proposed value here>*." (Use the matrix in Table 3.1). Say, "We've researched the tool and found it aligns to the customer's needs *<insert findings here>*. However, it is largely unproven and doesn't have the track record that our traditional tools have. What is your perspective on using it on this project?" An owner's response may often surprise a project team. Keep in mind, construction consumers are often keenly aware of the problems and gaps current systems have, and the openness to project funded R&D is sometimes a welcome approach to addressing and mitigating issues.

Ultimately, it's best to be honest with yourself and your team when putting together a strategy. In short: Identify your differentiators. Focus on real-world results. Drive at project-focused solutions. Aim to select the right tools for your project that create value. Accept that each project approach will require custom offerings. Align to your customers' goals. This winning strategy will create a much more meaningful dialogue with your customer beyond the traditional time and budget measuring sticks that drive the "construction as a commodity" marketplace (Figure 3.6).

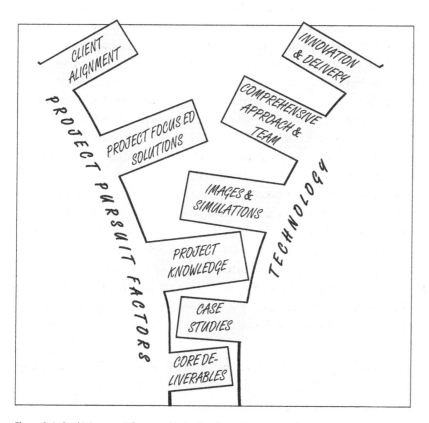

Figure 3.6 Combining pursuit factors and technology for a winning approach

Using BIM to Enhance the Proposal

When crafting an RFP response, it is important to both exhibit your firms' proficiency in using BIM and explain how BIM was used to create value. Owners see a lot of proposals that highlight the depth of a firm's experience and their capabilities in the areas specified by an owner on a particular project. Increasingly important is the ability to develop a comprehensive technology strategy and action plan that further strengthens a firm's position in the pursuit phase and reinforces their ability to exceed prior project performance.

Addressing BIM in the RFP

Many owners have specific requirements that they require submitting teams to deliver. There are now many examples of owners with clearly defined and published BIM requirements that factor into their project selection process. They include the General Services Administration (GSA), Los Angeles Community College, U.S. Army Corps of Engineers (USACE), and the State of Ohio, to name a few. Though these are just some of the owners requiring prescriptive BIM deliverables on their projects, it is interesting to note that some of the owner community is early in their technology adoption curve and may still be developing their requirements. Additionally, some well-informed owners, such as Dave Umstot (see the sidebar "Firsthand Owner's Experience and Vision of BIM" earlier in this chapter), say this allows teams to propose technologies that are project focused and custom built. Umstot says, "We developed empowering, rather than prescriptive, BIM standards to enable teams to enhance the value proposition and generation." This type of BIM criteria pushes back on design and construction teams to look at the project holistically and come to the table with a well-thought-out approach and toolkit.

Market trends of some amount of "real BIM" data being transferred to the owner and integrated platforms are becoming the new industry norm. This is an exciting opportunity for construction managers who understand the potential opportunity to deliver information in a format better than their peers, thus giving them a competitive advantage. In many cases, the opportunity for a construction management firm to become a "customer liaison" or "trusted advisor" in the BIM and technology space for an owner is a welcome offering and gives that firm the opportunity to better understand their customers' needs as well as have a deeper understanding of that particular client's pain points and how they can be addressed. A comprehensive list of available BIM guides is located on the buildingSMART website (http://bimguides.vtreem.com/bin/view/BIMGuides/Guidelines).

In the proposal stage, BIM and technology can be used to show capabilities in a number of ways:

- Images to illustrate the level of detail, clarity, or accuracy of the firm's modeling capabilities
- Simulations, such as animated videos showing the sequence of elements in a 4D schedule connected model environment

- Model-derived estimates, where the introduction of cost has been integrated into the model environment (5D), in some cases with the addition of schedule information as well

- Custom applications that demonstrate a firm's technological ability to manage construction, provide better clarity or understand items such as inventory, staff, or safety issues. These applications are often unique to firms

- Websites, tools, and portals used as a way to illustrate potential approaches to manage large infrastructure projects, or where community or customer engagement is needed

- Prefabrication scenarios and advanced work packaging simulations and coordination

- Site safety planning and community awareness visualizations (Figure 3.7)

IMAGE COURTESY: McCARTHY BUILDING COMPANIES

Figure 3.7 Site safety visualization

The use of BIM during marketing and pursuit efforts allows for more dynamic discussions than 2D plans do because the model can be used to illustrate strategies and components of construction in a more intuitive way. There are always new and innovative ways to use models during marketing and pursuit efforts. I encourage teams to use project pursuit opportunities to try out new technologies to better illustrate a construction process because of the fast nature of the pursuit. When you consider that some pursuits include the schedule, estimate, design and approach as deliverables, it is almost as if there is a mini-construction project to test out new ways of using BIM and technology.

Project Pursuit Images

Well-placed images in an RFP or interview can be invaluable. As the saying goes, "A picture is worth a thousand words," and the same is true in pursuing work and demonstrating ability. Adding BIM-derived images to a proposal can show a firm's expertise with a technology and a deeper understanding of the project, as well as more clearly illustrate a point or detail that is important. In turn, this limits the page count (often appreciated by owners reviewing the proposal) and breaks the monotony of text-only-based content. Additionally, an image or illustration is incredibly valuable in a presentation when describing a scenario or thought process. Images in presentations are preferred over bullet points and give the review panel something to discuss and explore further than text in question-and-answer sessions.

The following are a collection of images that illustrate various uses for BIM-derived images in a presentation. It is important to keep in mind that images can't "disguise" genuine study and research on how a project is to be built, the constraints, the opportunities, or the conditions. In fact, BIM-derived images in a presentation often highlight to what extent a pursuit team investigated various aspects of a project. For example, Figure 3.8 shows how a site logistics plan may be laid out for an occupied campus in an urban environment. Figure 3.9 shows graphically how stormwater will flow on a construction site and where protection needs to be added to prevent runoff. The 3D aspect adds clarity and dimension to items like the crane swing and height, and how traffic will maneuver on-site around construction conditions. These site logistics plans are useful in an interview and are an example of how a deeper amount of thought has gone into how a project site is laid out.

In the next example, Figure 3.10 illustrates a finished project from a specific view. Though these images are impressive, they should be used with the understanding that sometimes they aren't received well if the design team is separate from the contractor—mainly because architects typically view the design and aesthetics as their territory. Although it's fine to illustrate capability, be strategic about what value this may create from the contractor's perspective.

Figures 3.8–3.10 demonstrate that simple 3D or BIM-derived images can be used for great effect. Though using images may seem overly simplified, it removes the need to show a series of plan files illustrating a sequence or trying to depict an approach in a narrative. Remember, one of the values of BIM is the ability to visualize and communicate more effectively. To that end, a simpler approach with graphics showing what needs to be done can be more approachable and elegant than a more complex one.

Figure 3.8 Site logistics plan on occupied campus

Figure 3.9 Stormwater runoff prevention plan

Figure 3.10 Project pursuit rendering

Project Simulations

Simulations take static images one step further through the use of video to illustrate various parts of a project. Simulations for construction can take many forms and show different amounts of information. For example, to illustrate the construction sequence of a project you may use tools such as Navisworks, Synchro, Vico, or ConstructSim, among others, to show the combination of model elements being constructed over time. This layering, or "linking," of schedule information with model components is often referred to as 4D, or schedule simulations (Figure 3.11). These simulations are useful tools to clearly illustrate sequence and workflow and help create a clearer picture of how a project may come together. This is particularly valuable for customers who may not "think in 3D"; it is difficult for them to read multiple plans to understand the workflow. Additionally, these 4D simulations can be used to test project schedule data in the pursuit phase to ensure that the schedule is realistic.

Figure 3.11 Construction simulation QR code video

VIDEO COURTESY: AUTODESK

The rigor of these simulations can vary. Some simulations may show a more conceptual concept of construction, whereas others are directly linked to more complex and robust schedule information. The use of 4D simulations for marketing pursuits should be looked at from two perspectives. The first perspective is, what do you have time to complete? Understand that building these construction simulations takes time and requires the construction of a virtual model, at a level of detail that translates to the schedule (such as floor slabs broken into individual pours, or exterior walls broken up as they would be framed) as well as the creation of the schedule from which the model will "build" in the simulation environment.

Very simple 4D simulations can be created in a matter of days, but more robust and detailed simulations can take weeks or months to complete. The second consideration when creating 4D marketing simulations is, what are you trying to illustrate? Is the site very small and would it be good to show how equipment and materials will be unloaded and constructed? Is there a unique constraint for the project that needs to be brought to attention, such as safety control efforts, for an addition to an occupied hospital that needs to remain operational? It is important to ask if there are specialized circumstances that could be shown in the 4D simulations, which limits questions and can save time in a presentation by virtually "showing" how a particular issue may be addressed.

In today's construction market, some companies are making a best practice out of enhanced visualization and are hiring in-house specialists who create simulations, environments and videos for customers. Some of these specialists use rendering and gaming engines such as Lumion, Unity, 3ds Max, Autodesk Maya, and others that allow a participant to be "immersed" into an environment. Once in these environments, stakeholders can navigate on their own using gaming controllers, steering wheels, or other interfaces to explore their environment as they like. These simulations can also include scenario analysis. In simulations such as the Alaskan Way Viaduct earthquake simulation video completed by Parsons Brinckerhoff, BIM and rendering applications can go so far as to show disaster preparedness solutions in design and construction (Figure 3.12 and Figure 3.13).

Simulations continue to become one of the main drivers for the use of BIM through enhanced visualization. The simulation deliverable allows complex concepts to be displayed simply and clearly, which in turn gives customers the benefit of both appreciating the rigor and study as well as feeling more confident in a team's approach.

Project Pursuit Virtual/Augmented Reality Simulations

Virtual reality (VR) environments and augmented reality is a new way of experiencing construction projects. The cost of VR solutions has come down considerably and VR is increasingly being used as a means of immersing a user into an environment to better understand a concept. Users can "walk" around these environments using gaming controllers or hand controllers, or through head movement. The intuitive nature of these simulations allows customers to navigate to the spaces they wish to see and experience a project as they wish. These environments can be created through a number of tools, including gaming engines such as Lumion, EON, and Unity that fully render an environment, or "world." Other VR environments are prescriptive and can be rendered videos with preset paths built with tools such as 3ds Max, Maya, and SketchUp, among others.

Figure 3.12 Alaskan Way earthquake simulation

Figure 3.13 QR code link to video of simulation

Similar to virtual reality, augmented reality (AR) is the combination or layering of virtual information or objects on physical environments tools with reality. According to Wikipedia, AR is *"A live direct or indirect view of a physical, real-world environment whose elements are augmented (or supplemented) by computer-generated sensory input such as sound, video, graphics or GPS data. It is related to a more general concept called mediated reality in which a view of reality is modified (possibly even diminished rather than augmented) by a computer."* Examples of this technology are wearable technology such as Google Glass, Microsoft HoloLens, or Oculus Rift (Figure 3.14). These tools allow users to participate in two environments—the digital and the real—that can be controlled by actual physical movements or controls to provide an enhanced experience. Additionally, these tools may have potential impacts when marketing a project in areas of safety, job training, and project site management.

Figure 3.14 Oculus Rift Augmented Reality Headset

Other Marketing Tools

Other marketing technologies are becoming increasingly popular for teams pursuing work. Some of these are 3D printers, which you can use to print models of buildings, production plants, and landscapes at a relatively low cost. These 3D prints can also be used to show options and scenarios. The value in 3D prints is that they materialize the virtual environment and make it more tangible, which is particularly useful for a less technical customer. Additionally, applications have been developed for large-scale presentations and used as a way to organize the information required from an RFP and illustrate a firm's advanced use of technology.

Other offerings such as "smart" PDFs are beginning to take the place of traditional 2D PDFs. Smart PDFs can embed video content, 3D models, links to external websites, and links to other information within the document for ease of reference. Customers appreciate the ability to go into detail on certain topics of interest, whereas traditional PDFs miss the mark if information is not contained with the submission.

Tailor-Fit Your Offerings

In all of these potential solutions and deliverables it is important to focus on what will apply to a specific project and appeal to the customer (Figure 3.15). Construction consumers are not a "one-size fits all" bunch. Some responses may be more standardized than others, such as forms on contracts, insurance, bonding, and material standards. But when it comes to the ability to integrate creativity and use technologies to create an engaging response, the ROI is a sound investment of time and expense.

IMAGE COURTESY: LEIDOS ENGINEERING

Figure 3.15 Proposal project rendering

Often the difference between a winning response and a losing one is how well a team understands a project's unique conditions and constraints and is able to address them through enhanced visualization more effectively than their competitors. Using BIM as a means to communicate works, particularly when the customer better understands a concept—and even better when it allows a team to exhibit how they would work together on their project and what results should be anticipated.

Client Alignment

In his book *To Sell Is Human* (Riverhead Books, 2013), Daniel Pink explains the importance of "attunement" or customer alignment and describes attuning yourself to others by exiting your own perspective and entering theirs as an essential element to motivating others. In a value-based selection, the team that finds how to align to a client's goals and needs is often the team selected for the project. This is particularly true when it comes to the technology selected for a project. Decisions on the proposed tools steer the way a team collaborates and say a lot about how a team will work. When thinking about the tools to be used on a project, I ask strategic questions, such as the following:

- Does this BIM platform (Revit, AECOsim, ArchiCAD, Tekla, etc.) work with what the client has access to or has asked to be delivered?

- What expertise can be shown in the way a tool is used—something the client hasn't seen before?

- How are you planning on engaging the client and what tool will they be using to see their project's progress? Are they familiar with it? Does it work with what they are used to? Or is this an opportunity to show them something better?

- What would make this project run smoothly? Is there a system that you can introduce, consolidate to, or interconnect with that will create significant value?

- How do you envision working on this project? Do the tools you are selecting coincide with this plan?

Keep in mind that construction consumers are *smart* and whether you've researched a technology will be transparent to many clients. In fact, many are able to look at a construction management project proposal and quickly pick out which systems will work together and which won't. For this reason, think about the way technology is to be used and how it will benefit the customer and create value among the whole project team.

In her book *The Owner's Dilemma: Driving Success and Innovation in the Design and Construction Industry* (Greenway Communications, 2010), Barbara White Bryson summarizes working relationships as follows:

As with any relationship, (a) partnership must be built on trust and mutual understanding. Even within generally defined groupings of clients—institutional, commercial, non-profit—there exists a multitude of personalities, perspectives and expectations. Tapping into the potential that lies, sometimes admittedly well hidden, in the owner's heart and mind is a skill not to be undervalued. As technology continues to ease collaboration and speed information sharing, it is important to remember that clients are people too. Maintaining that human connection and engaging the myriad viewpoints that people bring to the process only enriches it. A truly cooperative and creative working process can pay great dividends for the ultimate success of the project.

Pushing the Envelope

Within a value-based selection effort, there is often the ability to "go big" and discuss or propose something innovative to a customer. For many BIM leaders, this is the fun part. Although it's important for a team to test a customer's temperature on innovative ideas and strategies, there have been many projects awarded to teams who weren't afraid to venture out and offer something new and different to a customer that placed them above the competition. Instead of spinning the model around in the presentation or talking about clash detection, the introduction of well-planned "new tech" can be a differentiator for a project team, especially if the proposed new technology addresses a customer's pain point in a new way that shows promise.

As mentioned earlier in this chapter, it is important to introduce new technologies, but the last thing a team wants to do is to propose a new tool and imply that they are experts in it, when in fact they just downloaded it this afternoon or got their first demo of the tool the day before. But what about a customer's processes? What if a team can propose a better way of working for them too?

In some cases, it is advantageous to question not only a technology, but the processes that are being used by a client to support it. This will give the pursuit team a better understanding of its use or perspective into what a customer values about it. If the project team can offer a unique solution and the customer is open to it, say something! Nothing is worse than a customer learning down the road that you knew of a better way to get something done but waited to talk about it. This type of open discussion is often appreciated by forward-thinking customers and shows a level of confidence and maturity on a project team that is often rewarded.

Seeking Value and Focusing on Results

This chapter has focused on how to go about winning projects with a cohesive technology strategy for value-selected projects. But what about price only or bid work? Can BIM play a part in providing value in this type of selection? Although the answer is yes, unfortunately the ability to optimize BIM's use and potential is limited. Because price is the driving factor in the hard bid or design-bid-build selection method, the ability to explore the potential uses, savings, and value in technology is mitigated. There may be contractual requirements to deliver BIM, but construction management firms will typically offer the "bare minimum" in order to stay as competitive as possible because cost is the motivating factor.

The information about a facility is becoming just as important as the facility itself. Why? Because for many facilities, access to building information, performance data and warranty logs can equate to millions of dollars a year in potential efficiency and risk. This will only continue to grow in significance and is often an opportunity for differentiation in a project pursuit. There are a number of ways to show how information is being developed into the model, analyzed, and shared among the project teams. As Sasha Reed puts it in "Defining BIM—What Do Owners Really Want?," "The goal of BIM is to tie together valuable information created, distributed and gathered during the project life cycle—ultimately, to remove inefficiencies in our processes and change the way we share, distribute and make use of information" (www.bdcnetwork.com/blog/defining-bim-%E2%80%93-what-do-owners-really-want#sthash.KksG5AxN.dpuf). In this explanation, having a comprehensive plan for handling a project's information is very valuable to a customer.

The modern-day construction market rewards the teams that don't focus solely on building a project to a schedule and budget. Rather, it is the teams that are finding new ways of creating value and alignment that are being rewarded in the construction industry. Just as architects who made the decision to stay the course with hand drafting instead of moving to CAD found themselves in a place that became increasingly difficult to compete as technology continued to change and the value discussion migrated to a more digital frontier, contractors who are not finding ways of using information from BIM and technology will find themselves set apart from their peers and unable to compete in anything but a price-driven commodity market.

The following guidelines have been proven by years of "painful" experience and lessons learned:

Don't propose too many new tools at once. Be realistic as to why you're proposing the tool and the desired outcome. If you introduce too many new tools at once, you may come across as trying to "reinvent the wheel" or trying to use the customer as a guinea pig. Remember, it only takes one more innovation than the competition to be viewed as "cutting edge."

Don't pick a tool based on it being flashy. Technology decisions should show clear value. Flashy tools only work once or twice with customers—value-added tools have a much longer shelf life.

Don't fake it.   The term "Hollywood BIM" caught on for a reason; it's best to save a reputation rather than put it at risk by promising unrealistic goals and mismanaging expectations.

Don't treat a customer's project as "just like that other project."   Unless that project went off perfectly and won numerous accolades, treat each project as unique and deserving of the right tools.

Don't think you know everything.   Aside from taking the fun out of it, pretending to be so experienced that there isn't any listening happening disengages customers and you miss the opportunity to develop more meaningful business relationships.

Choose and sell the right technology, not all of it.   Think about it like hiking. What do you need to get the job done effectively, without weighing down your ability to still hike up the mountain?

Select tools that have historically worked well.   There's something to be said for touting the proven advantages of certain tools, and they should be known.

Don't be boring.   It turns out clients are human. Be engaged and excited about the tools used and how they are going to deliver extraordinary value. It's hard not to like a team that has bought into an approach.

Innovate when possible.   The proposal stage is often a good place to show a firm's ability to adapt and show connectivity to market trends with new tools that can provide value. Use responsibly.

Align to the customer—always.   Look it may be the coolest thing ever, but always keep in mind it's the customer who pays for the project. Does that mean roll over and die if they don't like something? Not at all. Have the discussion and share the reasoning for why certain tools were selected based on their input. Often it is rewarding and appreciated.

Helpful Tips when Using BIM in a Presentation

The following are some tips you should consider before you make your presentation:

- Start your BIM computer up *before* walking into the meeting.
- Bring an extension cable.
- Bring a backup computer, adapters, and so forth.
- Practice live demonstrations exactly like you will be performing them (flying the model, live demos, videos, and so on).
- Don't have non-BIM users present on BIM. This is how you end up with clients thinking there are programs like "NavisCAD" and "Autoworks."

Summary

Marketing with BIM and technology can be a tremendously valuable tool in a pursuit team's arsenal. The ability to "show" how work will be done through images, simulations, and other technologies allows for unique opportunities to be explored with customers. Ultimately, the goal of using BIM for marketing efforts is to *show*, not talk, about a company's capabilities and leverage technology for enhanced deliverables.

A lot of tools are flashy and interesting, but focus on tools that create value, deepen the discussion, and create interest for the potential to use the project as a springboard for innovation. Remember that customers are humans too and are exposed to the pros and cons of construction every day. Offering technology solutions to a customer for a specific project in a clear, simple, and well-thought-out manner elevates a team from the crowd and can make the difference in winning work.

BIM and Preconstruction

This chapter explores the crux of the BIM process. This is where the job is set up for success and the value of BIM begins to reveal itself. We start by reflecting back on inspirational leaders in the building and manufacturing industry to learn from their approach to efficiency gains and technology adoption. We then delve into the critical path needs of project team setup, design scheduling, and practical workflows using BIM tools that will result in greater gains for construction and building performance.

In this chapter:

Leaning on the Past

In Henry Ford's book *Today and Tomorrow* (Doubleday, 1926) he says, "Efficiency is merely the doing of work in the best way you know rather than in the worst way. It is the taking of a trunk up a hill on a truck rather than on one's back. It is the training of the worker and the giving to him of power so that he may earn more and have more and live more comfortably." Henry Ford believed efficiency came from providing workers with the right tools so that they could work smarter, not harder. He didn't want them working all day and night to gain a livelihood, but instead aimed to arrange all the tools and machines in the most efficient way possible to simplify the process. He then empowered his employees to constantly improve the process by finding inefficiencies. They were to view everything as experimental.

This methodology evolved into the root of *lean manufacturing*, a term originally coined by John Krafcik in his paper "Triumph of the Lean Production System" (available at http://www.lean.org/downloads/MITSloan.pdf). It's the belief that when workers have a mind-set of constant improvement they will, by default, eliminate waste (time and material), improve process flow, and satisfy the customer. Today you will find this core idea, originally developed by the manufacturing industry, packaged up in various "lean" toolkits for the construction industry. These toolkits are designed to help the efficiencies of the design and construction process. Figure 4.1 shows some of the most popular methodologies out of McGraw Hill's 2013 *SmartMarket Report* on lean.

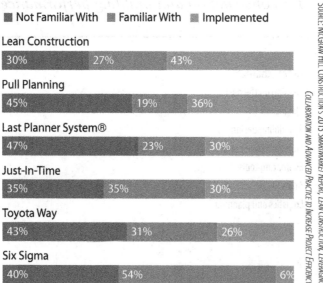

Familiarity With or Implementation of Lean Practices

Sources: McGraw Hill Construction, 2013

■ Not Familiar With   ■ Familiar With   ■ Implemented

Lean Construction
| 30% | 27% | 43% |

Pull Planning
| 45% | 19% | 36% |

Last Planner System®
| 47% | 23% | 30% |

Just-In-Time
| 35% | 35% | 30% |

Toyota Way
| 43% | 31% | 26% |

Six Sigma
| 40% | 54% | 6% |

SOURCE: MCGRAW HILL CONSTRUCTION'S 2013 *SMARTMARKET REPORT, LEAN CONSTRUCTION, LEVERAGING COLLABORATION AND ADVANCED PRACTICE TO INCREASE PROJECT EFFICIENCY*

Figure 4.1 A chart showing companies' familiarity with or implementation of lean practices

In 1997 Glenn Ballard and Greg Howell founded the Lean Construction Institute (LCI), whose mission is to educate the construction industry on these theories, principles, and techniques. Since the founding of the LCI, lean has become the buzzword for construction industry professionals, committees, groups, blogs, and forums around the world. Figure 4.1 shows that even with the term trending in popularity, there is still a good bit of educating to be done in the construction industry. What's interesting is that the concepts of lean construction are not new to our industry.

The Empire State Building

Let's go back to 1930 when one of America's most famous buildings was being built: the Empire State Building. Collaboration, prefabrication, 5S (Sort, Straighten, Shine, Standardize, Sustain), Six Sigma, value stream mapping (VSM), and just-in-time (JIT) delivery were all lean techniques that were used in 1930. Due to the successful implementation of these techniques, this monumental project was delivered early, under budget, and in an impressive time of only 13 months.

The Builders

Starrett Brothers & Eken were builders, entrepreneurs, and innovative thinkers. The Starrett brothers, Paul and William, alone had over 70 years of combined experience prior to walking in the door for the interview. The book *Empire State Building: The Making of a Landmark* by John Tauranac (Cornell University Press, 2014) recounts the interview for the project. Al Smith, former governor of New York and oversight for the construction, asked Paul Starrett how much equipment they had on hand to build the Empire State, and Starrett replied, "Not a blankety blank thing, not even a pick and shovel." He went on to tell them:

> Gentlemen, this building of yours is going to present unusual problems.
> Ordinary building equipment won't be worth a damn on it. We'll buy
> new stuff, fitted for the job, and at the end sell it and credit you with the
> difference. That's what we do on every big project. It costs less than renting
> secondhand stuff, and it's more efficient.

This honesty with Smith and John Raskob (one of the chief investors) earned their trust and they awarded Starrett Brothers and Eken the project after negotiating their $600,000 fixed fee to $500,000. Starrett gambled that other contractors would come in and sell the traditional approach, describing what they had and how they'd do it, but he knew the traditional way wouldn't yield the efficiencies required to meet a 14-month schedule. He believed that this project would require collaboration, innovation, and strategic planning to be a success.

Note: The findings out of the 2013 *SmartMarket Report* show a similar correlation to the adoption of lean construction and states that "in order for contractors to recognize the need to adopt lean practices, they must first recognize the inefficiencies inherent in a traditional" practice.

Collaboration

The Empire State Building was a fast-track schedule and Shreve, Lamb & Harmon (the architects) and Starrett Brothers & Eken worked as a team from the outset of the project. Carol Willis, director of the Skyscraper Museum in New York, said in a phone interview that the Starrett brothers were trained to be collaborators by their former employer, George A. Fuller Company, and that Fuller may have been the first contractor to embrace the integrated approach. This prior training and understanding played a major part in the success of the project. Willis writes about this trust-based relationship in *Building the Empire State* (W. W. Norton & Company, 2007) and cites separate quotes from the architect and builder that demonstrate a mutual respect for each other.

> *"Problems must be dealt with through authority greater than the architect possesses" to avoid "a duplication of effort and a loss of time too expensive to be tolerated."* – The Architect

R. H. Shreve, "The Empire State Building Organization,"
Architectural Forum 52 (1930)

> *"I doubt that there was ever a more harmonious combination than that which existed between owners, architects, and builder. We were in constant consultation with both of the others; all details of the building were gone over in advance and decided upon before incorporation in the plans."* – The Builder

Paul Starrett, *Changing the Skyline: An Autobiography*
(New York: McGraw Hill, 1938)

This approach, similar to that of Henry Ford, encouraged all members to work smarter and not harder by leveraging the collective mind. Everyone was in the boat and knew it, which led to some incredible innovations.

Innovations

Figure 4.2 shows one of the many innovative ideas that were used during construction. These rail systems were used on every floor and allowed for the efficient transportation of materials during construction, thus saving time and labor. The transportation of the common brick was a great example.

Some 10 million bricks were required for the exterior skin backing. Traditionally these bricks were dumped into the road and then manually loaded into wheelbarrows. Two wheelbarrows could fit on a hoist, each containing 50 common bricks. Starrett Brothers & Eken knew that this method was too inefficient to raise the 100,000 bricks required each day, so they had to create another method. They used the rail systems along with rocker cars to eliminate wasted time, motion, and effort as described in the following quote from the *Notes on the Construction of the Empire State* (Figure 4.3).

Figure 4.2 Industrial track (12 lb., 26-gauge rail) on 85th floor

It was no exaggeration to state that the bricks were untouched by human hands from the time they left the brickyard until the bricklayers picked them up to set down in place in the mortar.

Two brick hoppers were constructed in the First Basement. Each hopper had a capacity of about 20,000 bricks. Floor openings leading into these hoppers were made in the Main Floor. Trucks driving into the building dumped the bricks into these floor openings which were conveniently located near entrances to the building.

Each hopper fed the bricks through a slot opening into Koppel Double Side Rocker Dump Cars; each car having a capacity of about 400 common bricks. The loaded cars were pushed along industrial railway in First Basement and swung on turntables on to material hoists. After being raised

to proper floor, cars were pushed off hoist and sent along railway to point on floor where required for use by bricklayers in backing up the exterior limestone and metal trim.

Notes on the Construction of the Empire State

BUILDING THE EMPIRE STATE SKYSCRAPER MUSEUM ARCHIVE, IMAGE NUMBER: ES0132R.

Figure 4.3 A Koppel rocker dump car on the track to the brick hopper

Starrett Brothers & Eken calculated that using the rocker dump car saved them the labor of 38 men per day. The same method was used by Henry Ford at his Fordson plant. There were 85 miles of railroad to transport materials to any location in the plant. Raw materials would roll in one end and a fully powered Fordson tractor would roll out the other. In *Today and Tomorrow*, Ford wrote: "The thing is to keep everything in motion and take the work to the man and not the man to the work." These motion studies date even farther back to Frank Gilbreth and his Bricklaying studies, noted in his book *Bricklaying System* published in 1909. They all focus on eliminating waste in the time, motion, and effort of people and materials. Starrett Brothers & Eken thought outside the box and leveraged these manufacturing and building techniques on the Empire State Building, but this was only one of the innovative ideas to eliminate waste and keep production at peak performance. They also had two concrete mixing plants onsite in the basement to keep truck traffic down and built five temporary restaurants on different floors so workers didn't have to leave the site to go eat.

Planning and Prefabrication

Starrett Brothers & Eken also proved to be master orchestrators. William Starrett had a war analogy for planning that exudes the passion, effort, and thought that went into planning a project. He wrote the following prior to building the Empire State Building:

> *Building skyscrapers epitomizes the warfare and the accomplishment of our progressive civilization. Even the organization closely parallels the organization of a combatant army, for the building organization must be led by a fearless leader who knows the fight from the ground up, knows the hazards of deep foundation, and the equipment that raises the heavy steel and sets the massive stones one on another, the hoists and derricks, the mixers and chutes, the intricacies of all the complexity of trades that go to make up the completed structures; what they may be made to do, and where making must cease and daring must be curbed; where materials and things come from, and how long it takes to prepare the different kinds; what to allow for contingencies of temporary defeat, and how to consolidate the gains. Ever pressing forward, that leader with his lieutenants and they with their sub-lieutenants plan and do, ever prevailing over inertia, animate and inanimate, until the great operation fairly vibrates with the driving force of the strong personalities that direct the purpose of everything, seen and unseen, that makes for the swift completion of the work in hand.*

William Aiken Starrett, *Skyscrapers and the Men Who Build Them* (Charles Scribner's Sons, 1928)

The erection of steel on the Empire State Building demonstrated this understanding of planning, fabricating, sequencing, and scheduling. Due to the fast-track schedule and the limited space on-site, the coordination between the engineer, supplier, fabricators, deliverers, and erectors had to be seamless. Figure 4.4 shows the milestone schedule for Information Required, Mill Order, Fabrication Drawings, Delivery, and Erection. It also shows the tiers that each supplier was responsible for, designated by an "A" for American Bridge Company and an "M" for McClintic-Marshall Company. Lastly, it identifies the different cranes and their locations (A–H and K). Each piece of steel was labeled by crane, by tier, and by floor.

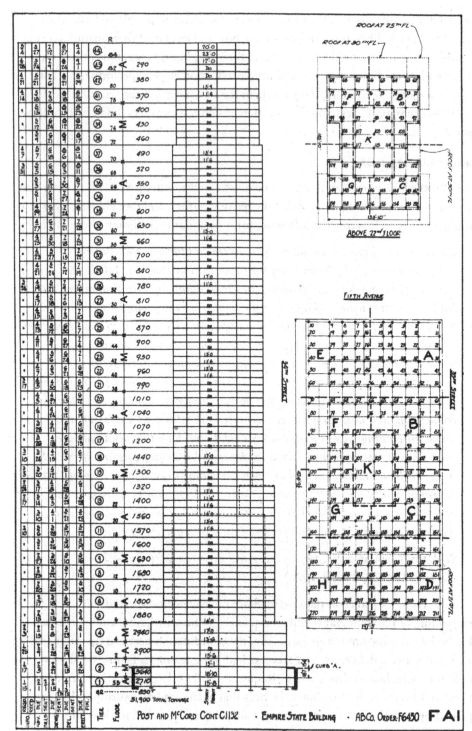

Figure 4.4 Post and McCord's fabrication and erection drawings

For the project, every piece was prefabricated off-site and uniquely designed to bolt or rivet into place. Once the steel was prefabricated and labeled, it was then sent to the jobsite "just-in-time" for it to be lifted and put into place. You can see how fast the Empire State Building was erected in Figure 4.5. They averaged 4~HF floors a week and lifted over 57,000 tons in less than 6 months!

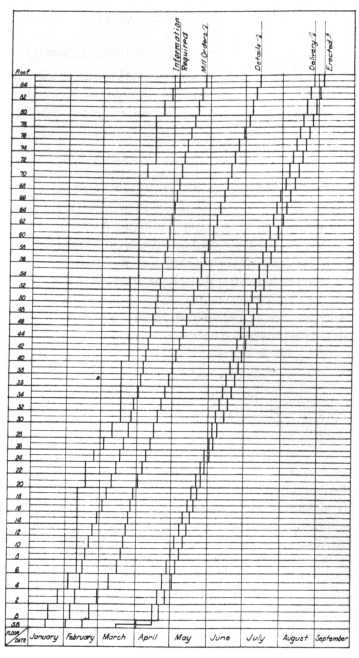

Figure 4.5 Steel schedule

Note: The time chart looks very similar to the Line of Balance Schedule (Figure 1.10) in Chapter 1, "Why Is Technology So Important to Construction Management?"

Again, the concept of just-in-time was a manufacturing technique that was used by Henry Ford. He knew exactly where every part was coming from and how long it would take to get to the manufacturing plant. In *Today and Tomorrow*, he wrote: "The shortage of a single kind of bolt would hold up the whole assembly at a branch." Ford boasted about the fact that his company didn't use or have a single warehouse. Ford as well as Starrett Brothers & Eken planned the work and implemented the plan using efficient scheduling, prefabrication, and just-in-time delivery.

An Empire State of Mind

Both companies believed in challenging the status quo, because they knew it wasn't good enough. They were creative inventors and innovators with a passion for building and manufacturing. They believed in relationships built on trust and synergy between team members; the idea that "the whole is greater than the sum of its parts." They approached every project with mastery level understanding and strategic planning, and would rather face defeat than be surprised. Some call it A3, Six Sigma, 5S, JIT, VSM, or the Toyota Way, but this was not a trend, a toolkit, or an acronym. This was a belief, a *behavior*, and a culture. It was an "Empire State of Mind" that strived to attack the project in the best way possible rather than the worst by leveraging collective experience and lessons learned. I believe Henry Ford said it best in *Today and Tomorrow*: "The only tradition we need bother about in industry is the tradition of good work." The Empire State Building was truly an inspiration to builders around the world and demonstrated a behavior of "good work" and operational excellence.

Adopting New Technology

The same behaviors that made the project team successful on the Empire State Building are available to us today. In addition, we now have BIM, which provides the ability to virtually construct a building in 3D. The question then becomes, what would someone like William or Paul Starrett be able to achieve with this technology? Could they have planned or built the Empire State Building faster than 13 months?

BIM is used synchronously with lean in reference to its ability to increase speed, efficiency, and reduce waste, but it's intriguing to look at how production of the construction industry has been trending over the last 70 years, since the construction of the Empire State Building in the early 1930s as well as the advent of CAD technologies in the 1950s and '60s (see Figure 4.6).

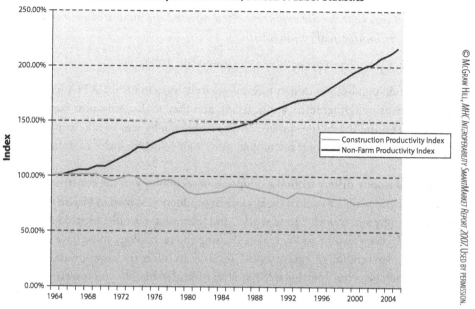

Constant $ of Contracts/Workhours of Hourly Workers
Sources: U.S. Dept. of Commerce, Bureau of Labor Statistics

Legend:
— Construction Productivity Index
— Non-Farm Productivity Index

Figure 4.6 Construction productivity index compared to nonfarm industries

So what's the cause of this trend? Shouldn't we be getting more efficient and faster as technology gets better? The answer to these lagging indicators can be found by "leaning on the past" of another successful company, Toyota.

No one can argue with the successes of Toyota and its Toyota Production System (TPS). Companies around the world strive to replicate their culture and process, which is based on those of Henry Ford. Their company, like Starrett Brothers & Eken and Ford, is centered around people and a "boots on the ground" approach. They believed in hard work, learning by trial and error, and getting your hands dirty, also known as *genchi genbutsu*. In his book *The Toyota Way* (McGraw-Hill Professional Publishing, 2003), author Jeffrey Liker recounts numerous conversations that reinforce a pride in building and an investment in their people. Unless something enhanced the building of cars, process, or people, it would not be considered for implementation in the TPS. What Toyota found is that often the best option was a simple solution:

> For example, in analyzing stamping dies that stamp out parts, the analysis technology (CAD) was not sophisticated enough to model the complexity of stamping out the part and verifying the best die design on a computer. So Toyota used a simpler solution that produced a color diagram showing the various stress points on the die. The die designer, working with an experienced die maker, then examined the diagram and made judgments on the design based on experience. In contrast, U.S. automakers implementing

CAD systems did this stress analysis using software alone, then made recommendations to the die designers in a throw-it-over-the-wall fashion. The result was that the die engineers often rejected the analysis because the results were impractical or unrealistic.

Jeffrey K. Liker, *The Toyota Way* (McGraw-Hill, 2003)

Toyota didn't embrace a design technology in its system until CATIA (computer-aided three-dimensional interactive application), and they took a long time customizing it before implementing it into the system. Toyota's philosophy was to take experienced professionals and surgically insert technology into their hands in order to optimize process.

That philosophy differs from the adoption of technology in the construction industry, and provides the answer to the decline in production shown in Figure 4.6. Toyota's philosophy starts with "good work" and technology is only adopted if it enhances it. That means the individual evaluating the technology must have a foundational understanding of "good work" in order to select the appropriate technology, which begs the questions: What level of leadership are you inserting BIM into your organization? How can people use tools (BIM) to enhance a process that they don't fully understand? BIM will provide limited value to any project, but if you want to build faster than 13 months, you need to look at the hands that are using the tools. I would hope they have some calluses. BIM will be only as good as the people who use it. Buying BIM software will not fix the inefficiencies in your organization, as you'll see in the next section. In Rex Miller's book *The Commercial Real Estate Revolution* he quotes a very intuitive formula, "OO + NT = EOO (Old Organization + New Technology = an Expensive Old Organization)."

The Journey to BIM

As you begin your journey of model management in preconstruction, focus and reflect on the lessons of these industry leaders. The advent of BIM is revolutionary and exciting because of how quickly you can manipulate the model with different concepts and ideas. However, this ability can be both a blessing and a curse. It's easy to become lackadaisical when using the capabilities of such a powerful software program and neglect the master builder skills required to build an Empire State Building. Parametric modeling enables designs to spiral out of control because of its efficiencies, which can create wasted time and effort during preconstruction if not managed properly. So how do you eliminate this waste in preconstruction? Refer to *The Toyota Way* and take an approach of *jidoka* (defined as "automation with a human touch)," which means building in quality as you produce the material.

From the 1930s until the advent of CAD, the Leroy Lettering tool was used to enhance the lettering process and exemplified automation with a human touch (Figure 4.7). Lettering was considered the most important phase of drawing and "a drawing could be ruined by careless and illegible lettering," according to authors Isaac Newton Carter and H. Loren Thompson in *Engineering Drawing: Practice and Theory, 2nd Edition* (International Textbook Co., 1943). Architectural and engineering drawings were about Pen weights, cleanliness of drawings, and artistic expressions. In the plans for the Empire State Building, every detail had to be carefully analyzed because of the amount of time that was put into every line and letter on a sheet. This effort created a certain amount of pride in the drawings and required experience, concentration, attention to detail, and practice. Creating a full set of construction documents involved highly skilled professionals.

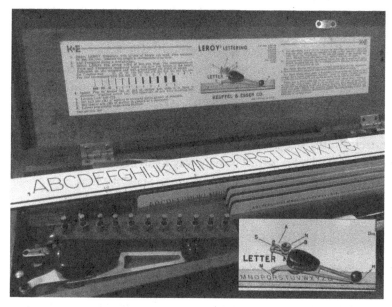

PHOTO COURTESY OF DAVE McCOOL

Figure 4.7 Leroy Lettering Tool

A common statement used to describe BIM claims, "Now we can build the building virtually before we build it." What's interesting about that statement is that in *Skyscrapers and the Men Who Build Them,* written in 1928, William Starrett described the design process, saying, "In fact, the building has, in a way, already been built on paper before the work we see starts." You see, they also built it before they built it, but there was more thought in the process. In a nostalgic way, it's kind of sad that this art of deep analysis and care is slowly disappearing.

BIM allows inexperienced users the ability to draw professional levels of lettering, details, isometrics, and renderings with minimal design and construction experience. The book *Engineering Drawing: Practice and Theory, 2nd Edition* gives a warning about this ability: "The novice usually goes to the detail drawings too soon, and many of the kinks of the general layout have not been straightened out, which may cause considerable re-design later of the details already worked out." This issue is only exacerbated by the automation of BIM and the fast-track schedule, which will be discussed later in this chapter.

Do you see how an efficient tool might lead to inefficiencies and waste if the process isn't defined? The Japanese people describe this kind of work as *muda*, or non-value-adding work. Having great tools like BIM doesn't necessarily mean you're free from *muda*; as you've read, it can unfortunately become a crutch for not doing things right the first time.

In our current state of construction delivery, we have reduced preconstruction efforts to a small fraction of the project compared to what we invested 100 years ago. We then wonder, "Why didn't everything come together? We used BIM." Sure, but did you take the time needed to fully understand the project, the required details, and as many issues as possible? Did you leverage experienced staff who have completed such projects before, such as the Starrett brothers? Preconstruction represents such a small portion of the construction budget that it is difficult to imagine a construction consumer who doesn't embrace the "go slow to go fast strategy" in lean that calls for a dedicated approach to project study in order to optimize the physical outcome. If we want to build like Starrett Brothers & Eken and change the trend shown in Figure 4.6 then we have to take an approach of *jidoka*. The only way to do that is "surgically" inserting BIM into experienced professionals with an Empire State of Mind.

The Kickoff

Congratulations! You've won the job and now you get to use BIM. You've developed a preliminary BIM execution plan, have a draft of a level of development (LOD) matrix, set up an FTP site with all your folders (toolbelt) organized, have a schematic site logistics plan that you developed for the proposal, and all the major designers and trade contractors (architect, structure, mechanical, electrical, plumbing, fire protection, and exterior) are on board. There are four keys to having a great kickoff:

- Get the right people in the room.
- Create the vision.
- Open the lines of communication.
- Remember to account for the *expectation bias*.

Getting the Right People in the Room

A common meeting that occurs at the beginning of construction phase is the *preconstruction* meeting, where the general contractor brings in the key team member's principals, VPs, and/or project managers to discuss the protocols for the project on a macro level. The topics include, for example, owner requirements, the pay application process, a tentative schedule, insurance requirements, and unique challenges on the job.

As construction progresses, general contractors will hold more micro-level meetings, called *preinstallation* meetings, at the jobsite with the field superintendents and tradespeople. The purpose of these meetings is to review the scope of work that the trade contractors will be performing on the project down to how many nails are required in the roof shingles.

When building virtually, you'll have a similar meeting before modeling begins, commonly referred to as the *BIM kickoff* meeting. It's a hybrid meeting, because you'll discuss both macro and micro topics on how to build the building virtually. This meeting is critical to the success of the project and requires project managers as well as the virtual project manager (VPM). Keep in mind that not having the VPM in the room for this meeting will guarantee failure out of the gate. The sidebar "Kicking Off to a Bad Start" walks you through a scenario.

Kicking Off to a Bad Start

BIM manager: "Where's your VPM, Jill (mechanical PM)?"

Jill: "Oh, he couldn't make it, cause he's wrapping up this school we're working on. I'm going to take notes and relay the details to him."

BIM manager: "Did you not get my e-mail that said that it's mandatory to bring all lead VPMs?"

Jill: "Yes, I apologize, but we're slammed right now. I'll make sure he gets up to speed with everything we discuss today."

[The BIM manager goes through introductions and begins the kickoff meeting. He opens with a vision for the project and begins to discuss the importance of communication/collaboration. The PM seems to be following along quite well and can relate to the importance of both topics. Eventually the BIM manager gets into the uses, model creation, and how the information will be exchanged among the team members (information exchange plan).]

BIM manager: "All right, so everyone using CAD needs to be saving down DWGs to 2010 for Frank. Make sure you save those in the CAD folder and not the Native folder. John, you'll only need to take your model to LOD 300 for those boom supports, but Jill, your detailer will need to take the hangers to 400 in order to create the CSV for the total station. We're going to be using Tekla BIMsight for coordination. If you're using Revit, then you'll need to export to IFC for BIMsight but export to DWG for the CAD users. Don't forget to save down to 2010. Your RVTs will still be saved in 2015. Susan, you'll need to get Jill's lead detailer the GBXML files for the preliminary building energy model."

What do you think those notes will look like when they reach the VPM? I've always wanted to know, but I can only imagine that it's like the Game of Telephone, where you whisper in someone's ear and they pass it around until the last person announces something completely different from the original message. The BIM kickoff is the kickoff for installing the work virtually and should be viewed as such. The individual installing the work must be there.

Creating the Vision

The BIM kickoff is one of my favorite meetings. You get to meet with new and old friends who are like-minded people. It's a fresh start and a clean slate. You can forget about any pitfalls you might have had on the previous project and bring all the lessons learned to this one. It sets the tone for the project, and it's where you get to present all of the hard work you've put into the plan. But most importantly, you get to create the vision for the project.

In *Leading Change* (Harvard Business Review Press, 2012), author John Kotter discusses the importance of creating a vision within a team or coalition. Kotter says a vision serves three important purposes:

- It clarifies the direction.
- It motivates.
- It helps coordinate the actions of different people.

These are not easy tasks. Managers have a tendency to tell people what to do without getting buy-in, which creates an "I win/you lose" philosophy. They tend to focus on the facts instead of discussing the purpose of why we're doing what we're doing. The vision should be simple in order for people to believe and understand it. Here are two examples:

Vision #1   The schedule for coordination is 5 months, but by leveraging the parametric abilities of BIM and by collaborating through a federated model in Tekla BIMsight, we'll be able to create efficiencies in both document creation and coordination. The goal of coordination will be to identify constructability issues early, which will decrease the number of RFIs and potential change orders. We will also be able to quickly analyze the building's orientation and materials to calculate loads in order to properly size the equipment for the job, thus reducing energy consumption and providing more daylight to the end users.

Vision #2   This job will be fast and it will be a challenging. It will require you to trust and rely on each other. If we stick together, then it will be one of the most profitable jobs you've been a part of. At the end of the day, we will deliver a unique high-performance building that everyone can be proud of and future generations will thank you for years to come. Are you ready?

Which vision do you think has more impact? Stephen Covey refers to Vision #2 in *The 7 Habits of Highly Effective People* as a win/win philosophy, where "it's not your way or my way, it's a *better* way, a higher way." It hits at the core of why we do what we do. We're all in the boat and we win or lose together. If you are the VPM, your work is not over once you've created this vision. You have to believe it, you have to repeat it, and you have to exemplify it. If the team can't rely on you, you'll lose their focus. If team members aren't reminded of the vision during the challenges, then they'll falter. If you don't lead by example, they will follow your path. Telling your team members what to do will never be as effective as creating a vision that they'll want to do.

Opening the Lines of Communication

Do not talk the whole time and don't get too attached to your execution plan! The BIM kickoff should be an opportunity to share each other's lessons learned. As we've said, technology moves too fast for you to keep up with, so the only way you're going to be able to sift through all the programs and add-ins is by collaborating with the other like-minded BIMers. Your execution plan and best practices should be based around solid processes, so that if someone brings an idea to the table you can quickly assess if it enhances your people or process. The FTP sites and file management programs are a perfect example.

    The way in which we share information will get faster and more automated. Your folder structure and filenaming might not change, but someone may recommend a new FTP or file-sharing client that has better sync functions or speed that allow you to be more efficient as a group. That's a quick switch and doesn't have any effect on your process. It only makes your team better. Use the BIM kickoff as an open forum for collaboration and ideas and then modify your execution plan as needed.

Accounting for the Expectation Bias

Remember Craig Dubler's quote that "BIM is only as good as the least technical person." The least technical people will nod their head during your kickoff, and when you ask if there are any questions, will remain silent. Imagine what it's like for them sitting in the room with a bunch of savvy BIM gurus. It can be intimidating. Have you ever been in a class where you have a question but you don't want to sound naive in front of your peers? We've all been there. What's interesting is that the teachers almost always know which students get it and which ones don't. The difference is that teachers know that their students, for the most part, are there to learn, which gives them intuition in knowing when students are struggling to understand.

    The BIM kickoff is not necessarily a classroom, so it's easy to get distracted by the excitement of a new job, the vision, the collaboration, and the execution without recognizing the one person who's actually there to learn. Don't forget to take into account the *expectation bias,* defined in Chapter 2, "Project Planning." When you're aware of it, you'll develop that intuition. Once you've identified those people, you can focus on bringing them up to speed so it doesn't affect the team.

Scheduling Design

The growing trend and use of fast-track integrated project delivery methods (IPD, DB, and CMaR, as described in Chapter 2) creates a complex coordination of design and construction information, because they now occur simultaneously (Figure 4.8).

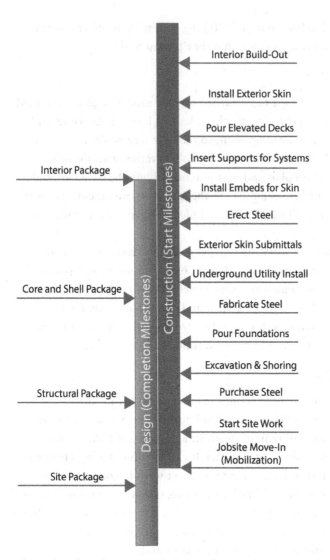

Figure 4.8 Design and construction schedule example

As discussed in Chapter 2, fast-track is accomplished by breaking up the design into design increments, shown on the left side of Figure 4.8, as opposed to waiting for the entire design to be complete before construction starts. This allows construction permits to be released in phases, which decreases the total design and construction duration and can reduce costs associated with material inflation and the general overhead. Let's break down Figure 4.8 to dig a little deeper into the relationships of these design increments.

Figure 4.9 demonstrates a four-increment design schedule with a one-year duration. Notice how the Design Development (DD) phase happens at different times in each increment, blurring the line of DD at any given point. If you go back to the

MacLeamy Curve, shown in Figure 2.3 in Chapter 2, you'll notice that the most valuable time to solve issues is during the Schematic Design (SD) and DD phase. Well, now there are four MacLeamy Curves, so in order to achieve the most value, you have to fully understand *what* information should be contained within each increment at DD. We'll look at an example from the AIA of the information that is typically contained in a DD phase.

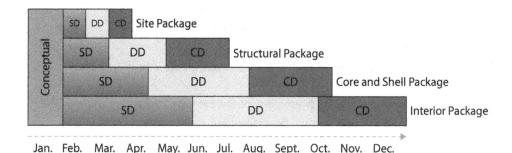

Figure 4.9 Increment schedule

In 2011 the AIA released a document called the *Design Development Quality Management Phase Checklist* to help teams determine what information was required at the DD phase. You can find it at http://www.aia.org/aiaucmp/groups/secure/ documents/pdf/aiab094998.pdf. This document is a good starting point to help you understand what information is required in a linear design with only one increment, like you'd find in a Design-Bid-Build. It is also a good place for us to begin to demonstrate how fast-track becomes a strategic sorting exercise of information.

The AIA document contains five major checklists:

- Civil/Site Design
- Design/Architecture
- Structural
- MEP/FP
- Other Consultants

Each checklist contains 7–29 items with check boxes down the left-hand side. Every item in the column is a minimum design criterion that the AIA has determined to be necessary for the completion of the DD phase. Your team would read the items and "check the boxes" if the item was complete. Once all boxes were checked, you would confirm that your design development phase was complete. We'll take the "MEP/FP" checklist of the document to demonstrate how information sorts out among the various increments (see Table 4.1).

Note: I've modified Table 4.1 slightly with counts instead of check boxes and changed the "Notes" column to "Increment." I then indicated which design increment I thought the information belonged in.

▶ **Table 4.1** Design development quality management phase checklist for MEP/FP

	MEP/FP	Increment:
1	Design criteria including indoor and outdoor conditions, ventilation, air circulation, minimum exhaust, sound levels, system diversities, and building envelope thermal characteristics reaffirmed in the documents.	Core and Shell
2	All riser diagrams are complete.	Core and Shell
3	Typical floor(s) coordination is complete with all risers, chases, and interstitial ceiling areas coordinated.	Interior
4	Major equipment rooms are laid out and final space requirements are confirmed.	Core and Shell
5	Plans other than the typical floor are sufficiently developed, giving a reasonable expectation that final coordination will not adversely impact the architectural layout, structural design, and so forth.	Structural
6	Mechanical floor plans should be nearly graphically complete with double-line ductwork.	Core and Shell
7	Plumbing floor plans should be nearly graphically complete showing horizontal collection and distribution piping.	Core and Shell
8	Coordinate mechanical, fire protection, and plumbing risers with the floor plans.	Interior
9	Coordinate horizontal distribution of major ductwork, plumbing piping, sprinkler mains, etc. with architectural reflected ceiling plans such that the ceiling heights can be confirmed. Locate diffusers, light fixtures, and other principal devices.	Interior
10	There should be enough coordination between the various MEP/FP disciplines to confirm interstitial spaces and large structural openings in slabs and shear walls.	Structural
11	Equipment cuts that are exposed to view in public areas are finalized.	Interior
12	For projects where DD is a preliminary guaranteed maximum price (GMP) issue, all equipment schedules are completed and a draft technical specification needs to be prepared.	Core and Shell
13	Reaffirm energy code analysis.	Core and Shell
14	Coordinate utility requirements.	Site

Do you see how the information spans across the different increments? The increment selection was my opinion, but this same analysis could be done with the design team for each of the five checklists and would produce a solid foundation of what information is required to complete the DD phase of the four design increments. However, understanding what information is in each increment is the easy part of the problem, and stopping here is where many Design-Build teams start to lose control of the design organization and schedule. The understanding of what information is required in the increments is the first part of solving what I call the *incremental dilemma*. The second, and most challenging, part is resolving *when* information is required.

Creativity and design are dynamic in nature and involve iterations or cycles to get to conclusions.

"That didn't work well, so let's try this one."

"What if you did it this way?"

"Don't change your design. Let me see what I can do on my end."

"Can we look at a more sustainable product?"

These conversations are common during design and are necessary in achieving a high-performance building. The creative process of design requires an "ebb and flow" of ideas. In traditional scheduling like the Critical Path Method (CPM), commonly used in construction, one activity ends, which allows another activity to begin. It is a linear process that doesn't allow for iterations or interdependence, so scheduling "when" design information is due is problematic because it doesn't take into account the cyclical process. The problem is that CPM is the most common method used for scheduling the design. If you went through all the checklists in the AIA document, determined what information was required for the structural package, and then drew dependency arrows between the activities, you'd end up with something similar to Figure 4.10. The mechanical duct increased in size and requires larger openings in the floors ... which causes the structural framing to change, causing the interior walls to shift ... which changes the ceiling layout and now causes the lights to adjust because they're not symmetrical ... and all of it was because the architect changed the R-value on the exterior skin in the Core and Shell package, which changed the loads for the mechanical systems. It's a "butterfly effect," or "chaos theory." The design process is based on these interdependencies and thus creates a scenario where cyclical patterns can be seen everywhere, but no one knows what activity comes first. If you handed the structural engineer all the "what's" out of the AIA document and said, "All of this needs to be resolved by the middle of May for your DD package," you can imagine what would happen.

The increments complicate the "when" even further because now the interdependencies are stretched over 8 months for DD as opposed to the single linear design where DD may have only lasted 3–4 months. What that means is that specific information of later increments must be accelerated to finalize the earlier increments, but now there's 8 months of sorting to figure out what information to accelerate. There are three possible scenarios that occur when information isn't delivered in time for an increment:

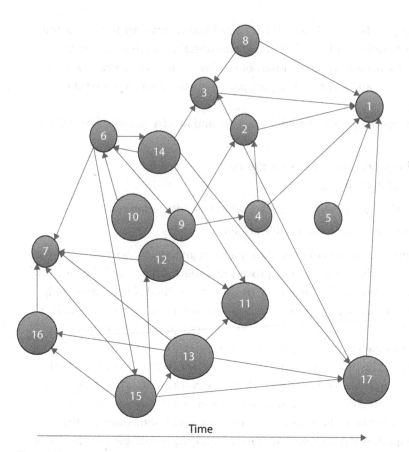

Figure 4.10 Design information chaos

Scenario 1   The first scenario involves taking advantage of the permit review process. Typically an increment package will go in for a permit review and the governing authority will return it with comments that need to be resolved for a second-round resubmission or "backcheck." Some design teams count on the backcheck as additional design time to sort through missing information, make final adjustments to the increment, and slip the changes into the resubmission. This is less than ideal and can lead to another backcheck, which could delay the construction.

Scenario 2   The second scenario involves receiving information from a later increment that affects an earlier increment that has already been permitted. This results in an addendum to the package that has already been issued for construction. If the construction team has already started construction, these addenda can have significant impacts on the costs of a project.

Here's an example: The structural package has been permitted and the foundation walls and footings are being poured. However, the information for the elevator is being coordinated in the Core and Shell package and the design team realized during the plan review that the elevator pit won't work with the specified elevator manufacturer. This issue results in a change to the foundations and walls that have already been poured, because the structural engineer was forced to make assumptions on the elevator in the structural package prior to having all the information.

Scenario 3 The last scenario is caused by not realizing the missing information until the exact moment when it's required for construction. This results in the RFI process that was discussed in Chapter 2, and completely contradicts the goal of Integrated Project Delivery methods.

At this point, you may be wondering what this incremental dilemma has to do with BIM. I wish I could tell you how many times I've heard architects, engineers, and subcontractors say, "I'm way over on my detailing budget." You may be reading this book because of that statement. One plausible cause is due to the learning curve associated with BIM, but I'm hearing this from industry leaders. If it's not the learning curve, maybe it's just bad estimating, but is everyone really that bad at estimating BIM? I don't think so. What I've found is that these industry leaders are blowing the budget because of the incremental dilemma. The teams don't have a clear understanding of "what" is in each increment and the "when" is driven by the interdependencies of the "what's." They end up trying to catch the "butterfly effect" with BIM, which is a bigger challenge than they thought it'd be.

The dilemma of design organization is not easy, and institutions like MIT, Loughborough University, Lockheed Martin Tactical Aircraft Systems, the Department of Mechanical and Aerospace Engineering, and NASA have all been researching this issue for years. The solution to the incremental dilemma has been around since the 1960s but was officially coined by Don Steward in 1981. It's known as the Design (or Dependency) Structure Matrix (DSM).

Design Structure Matrix

The DSM is structured like a spreadsheet and uses an algorithm to optimize the iterations of the design process based on information dependencies. You can do the calculations manually, but I prefer computer automation. You'll find a few DSM software programs at www.dsmweb.org, but to demonstrate the use of a DSM, we'll use the free DSM Matrix provided by ProjectDSM, which you can download at www.projectdsm.com/Products/Downloads.aspx. The DSM Matrix is a software program designed to help individuals understand the fundamental concepts. To use it, perform the following steps:

1. Download and install DSM Matrix.

2. Open the DSM Matrix.

3. Press Ctrl+N to start a new project.

4. Use the Name column to add items from the AIA checklist that you feel impact the Structural package (see Figure 4.11).

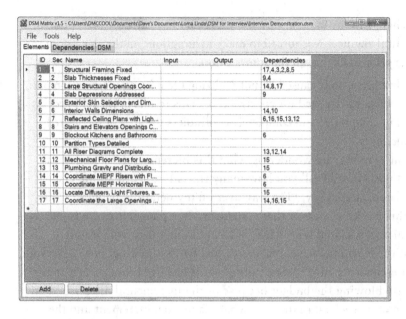

Figure 4.11 DSM elements

5. Click the Dependencies tab.

6. Start at the top of the list in the Selected Reference Element window and select your first element.

7. Use the Add/Remove Selected Source Element button to add elements that your selected reference element is dependent on (see Figure 4.12).

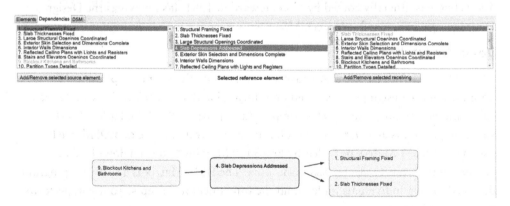

Figure 4.12 Design process mapping

8. Continue this process until all element dependencies are complete.

Click the DSM tab, and you'll see that all the elements have been automatically listed and assigned a number vertically and horizontally in the chart (Figure 4.13). The dots in the grids represent dependencies. The dots to the right of the diagonal line represent potential "butterfly effects," or design iterations.

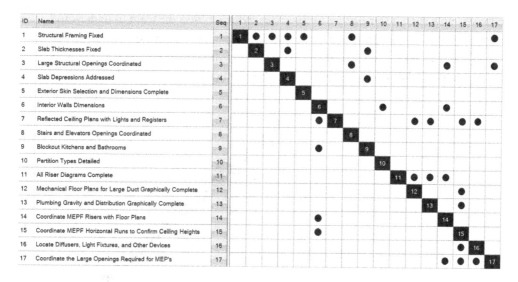

Figure 4.13 DSM before dependency sequence

For example, element 1, Structural Framing Fixed, is dependent on element 17, Coordinate the Large Openings Required for MEPs. If you scheduled using a CPM in this exact sequence, 1–17, then the structural engineer would be waiting from 9–16 for information that could potentially impact framing. That's a long iteration and creates wasted effort. This is where the algorithm comes into play. The algorithm looks at the dependencies in order to optimize the iterations.

9. From the Tools menu, select Dependency Sequence to optimize the iterations (see Figure 4.14).

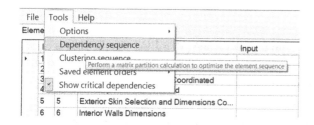

Figure 4.14 Selecting Dependency Sequence from the Tools menu

The software now has optimized the iterations of design cycles to make the process more efficient (see Figure 4.15). Not only has it optimized the cycles, but it has identified where healthy iteration will exist (Interior Wall Dimensions and Coordinate MEPF Risers with Floor Plans). This gives your team an opportunity to accelerate, innovate, or estimate the information. You could *accelerate* the MEPF riser information to finalize the wall dimensions. You could *innovate* the MEPF risers in an effort to eliminate their interdependencies. Lastly, you could *estimate* the MEPF risers and overengineer wall thicknesses or shaft sizes, for example. Whichever option your team chooses, now there's some order to the chaos, which allows the team to understand where the iterations exist in order to make decisions and schedule the design. The best part about this for BIM is that it gives the detailers direction on modeling and when it's appropriate to take the model to the next LOD, which helps them catch the butterfly.

ID	Name	Seq	1	2	3	4	5	6	7	8	9	10	11	12	13	14	15	16	17
5	Exterior Skin Selection and Dimensions Complete	1	5																
8	Stairs and Elevators Openings Coordinated	2		8															
10	Partition Types Detailed	3			10														
6	Interior Walls Dimensions	4			●	6													
14	Coordinate MEPF Risers with Floor Plans	5				●	14												
9	Blockout Kitchens and Bathrooms	6				●		9											
15	Coordinate MEPF Horizontal Runs to Confirm Ceiling Heights	7				●			15										
4	Slab Depressions Addressed	8						●		4									
12	Mechanical Floor Plans for Large Duct Graphically Complete	9								●	12								
13	Plumbing Gravity and Distribution Graphically Complete	10								●		13							
16	Locate Diffusers, Light Fixtures, and Other Devices	11								●			16						
2	Slab Thicknesses Fixed	12						●		●				2					
7	Reflected Ceiling Plans with Lights and Registers	13				●				●	●	●	●		7				
11	All Riser Diagrams Complete	14					●				●	●				11			
17	Coordinate the Large Openings Required for MEP's	15					●			●			●				17		
3	Large Structural Openings Coordinated	16		●														3	
1	Structural Framing Fixed	17	●	●						●					●			●	1

Figure 4.15 DSM after dependency sequence

Note: This is a very basic example of a DSM. There are many variables to dependencies, and you should do further research before implementing a Design Structure Matrix on a project.

Scheduling the LOD

Perhaps the highest process hurdle to jump over is the idea that BIM equals what is to be built. As discussed earlier, a model doesn't need to three-dimensionally represent every last doorknob and hinge, because doing so is neither productive nor efficient; rather, the model should contain enough information so it can be constructed. By leveraging the DSM and the LOD matrix, a team can define what level is "enough" in order for productive collaboration and early decision making. According to Randy Deutsch, author of *BIM and Integrated Design: Strategies for Architectural Practice* (Wiley, 2011):

> The key is to identify and prioritize the kind and amount of information and detail required to (1) meet the goals and the expectations for the phase you are in and (2) make progress toward addressing later phases and team requirements.

If you compare Figure 4.9 to Figure 4.16, you'll notice LOD stacks similar to the development of design. LOD "300" is now blurred across the increments, which is why the BIMForum states that there is no such thing as a LOD <Insert level here> and set er also model. In the month of May, you have anywhere from 200 to 400 LOD in the model. This is where the power of the DSM is so compelling. By leveraging the DSM, you can control the acceleration of model content and manage the job more efficiently. Now you know when information is required at LOD 400 and when information can remain at LOD 200. You can also streamline coordination during design by prioritizing the building components and MEP systems efficiently (we'll discuss this topic more in Chapter 5, "BIM and Construction"). Lastly, with organizational knowledge, your team can accurately budget for the project.

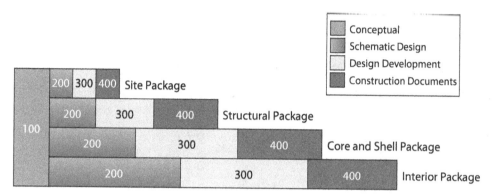

Figure 4.16 LOD schedule

The first step to solving the incremental dilemma is understanding the "what" and assigning the dependencies to determine the "when." This allows you to strategically plan the design as opposed to chasing it with BIM. Once all the elements are identified and prioritized, you can use a software program like Pull Plan (www.pullplan.com) to schedule from the construction milestones. This will help you determine the durations required for each iteration to meet the demands of the fast-track schedule.

> *The design manager needs to be crystal clear about the level of detail needed for each design submittal. For example, the detail needed for early permits may be less than that required for bid packages for equipment and subcontractors, or for the final construction documents.*
>
> Charles Pankow Foundation, *Design Management Guide for the Design-Build Environment, Version 1.0*

Constructability Review

Constructability review, means and methods, and *project construction feasibility* all refer to the evaluation of whether the design can actually be built by a construction team and how it will be done. Aligning to the promise of BIM, the purpose of models in the constructability review phase is to simulate and analyze the actual construction issues while they're cheap. It would be a disservice to assume that every job will use one of the Integrated Project Delivery methods, prioritize design information with a DSM, schedule with pull-planning, and model all trade content to LOD 400 in a live "real-time" consolidated or federated model. This is achievable, but it's not our current state, as seen in the *2014 SmartMarket Report* (Figure 4.17).

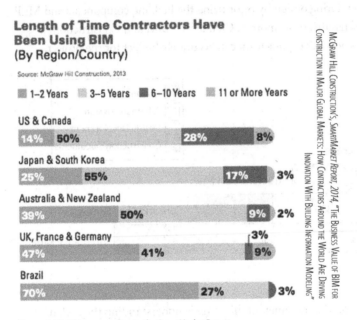

McGraw Hill Construction's, *SmartMarket Report, 2014*, "The Business Value of BIM for Construction in Major Global Markets: How Contractors Around the World Are Driving Innovation With Building Information Modeling"

Figure 4.17 BIM use according to the *SmartMarket Report*

Constructability review during design is all about the ability to see "beyond the clash," meaning not everything is solved with a click of a button. As we discussed in Chapter 2, design models typically show design intent and do not carry the level of development during the DD or CD phase that would be required for construction installation or fabrication. Until BIM authoring software creates LOD 400 out of the box, there will always be an evolution of detail contained within the model elements during design. Another reason for this is due to efficiency and speed. Architects and engineers like to keep their models on a diet during the design process. More information equals a heavier and slower model to manipulate in the authoring software. This will be less of an issue as computers get faster and we increasingly move toward cloud computing, but it's important to note. These two factors further reinforce the fact that in order to achieve the ideal MacLeamy Curve of cost savings, you will need to have an in-depth knowledge of construction installation and use a hybrid approach of 2D and 3D analysis in order to see "beyond the clash."

Leverage the Plans

It's amazing how the model has hypnotized experienced architects, engineers, contractors, and subcontractors. The transition to clash detection for constructability review since 2007 has been radical, with tools like Navisworks and Tekla BIMsight emerging. At first, people said, "You can't rely on the model. It won't be built like that." Now they say, "How can I coordinate if it's not in the model? It has to be in the model!" People have become so dependent on the model content that they have started neglecting the drawings. They will spend hours reviewing 3D models instead of spending 15 minutes reviewing plans. I'll give you an example from construction 101: *door framing and ductwork*. Figure 4.18 shows typical framing around a door. There are two king studs that go from floor to floor and a header across the top. These studs are referred to as "critical studs" because they have to be in that exact location to support the doorframe. They take priority over MEP systems, but MEP engineers are notorious for routing duct or system racks directly over these "critical studs" during design.

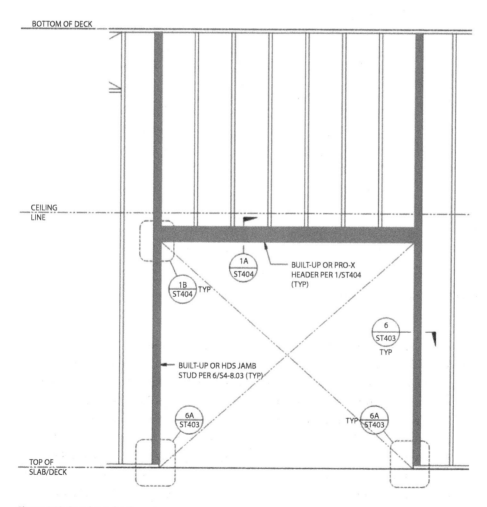

Figure 4.18 Door frame detail

The majority of architects do not model framing, so how do you identify a clash between something like ductwork with an element that doesn't exist? Well, you can't, but because of the hypnotizing effect of BIM, some may spend hours searching the 3D view of the model trying to find these conditions. What seems like a logical solution is to bring the framing subcontractor in early to model studs during the design. Often what ends up happening is the framing subcontractor tries to catch the butterfly, because the design program is still evolving, which results in *muda*. Even if clash detection were leveraged in this scenario, it would still take a considerable amount of time to look at each condition one by one. Now look at Figure 4.19 and Figure 4.20.

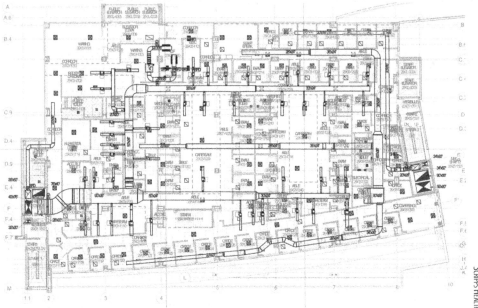

Figure 4.19 Mechanical plans before markup

Looking at the plans makes it much easier to identify potential issues with *critical studs*, which saves you a significant amount of time in design review and detailing. The plan views are produced from the model; therefore, they represent what is to be built. If it's a problem in the plans, it's going to be a problem in construction. Once you've identified the issues, you can talk through the solution with the mechanical engineer and the framing subcontractor to resolve them efficiently. This method may not require modeling every stud; it may simply require a construction mind-set and early collaboration. It's essentially going back to a light table review, and as professed in the *Toyota Way*, often the best option is a simple solution. The same process can be used for plumbing racks, electrical racks, and electrical fixtures.

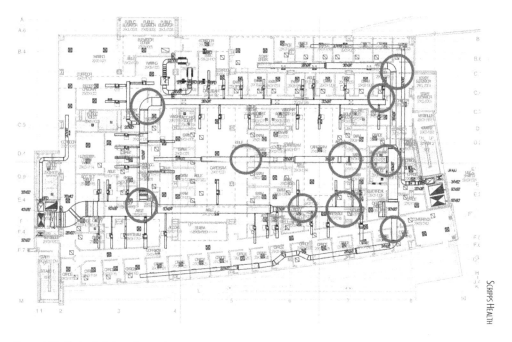

Figure 4.20 Mechanical plans after markup

Clash detection is an incredible tool, but common sense and readily available information can save significant time and allow team members to reach their coordination goal faster. In this sense, a balance between plan and 3D views are often the most viable option.

Leverage the Details

In Chapter 2 you read about the new project engineer who was surprised by the structural kickers in the field. The kicker was annotated in the drawings but wasn't in the model, which forced the ceilings to be lowered. That's a tough lesson to learn and you never want to disappoint a client, but that issue pales in comparison to the disappointment a client will have if you overlook water infiltration or fire-life safety. Those two items can be very hard to identify using clash detection software, but they are critical to the design and performance of a building. In this section, we will focus on water infiltration.

The National Roofing Contractors Association (NRCA), founded in 1886, "is one of the construction industry's most respected trade associations and the voice of roofing professionals and leading authority in the roofing industry for information, education, technology and advocacy." (http://www.nrca.net/About/) *The NRCA Roofing Manual: Membrane Roof Systems—2011* lists common details that are industry standards for constructing a roof and are typically referenced in design documents and roofing manufacturers' submittals. Figure 4.21 shows the minimum horizontal distance of separation between elements for waterproofing.

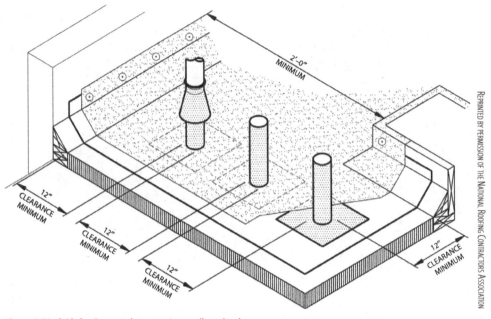

REPRINTED BY PERMISSION OF THE NATIONAL ROOFING CONTRACTORS ASSOCIATION

Figure 4.21 Guide for clearances between pipes, walls, and curbs

Figure 4.22 shows the minimum vertical distance that must be achieved for waterproofing.

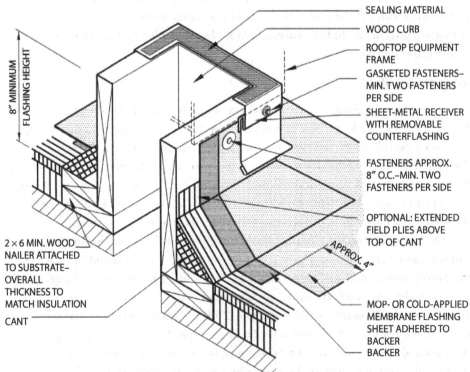

Figure 4.22 Base flashing at wood curb

REPRINTED BY PERMISSION OF THE NATIONAL ROOFING CONTRACTORS ASSOCIATION

Once you've reviewed both Figures 4.21 and 4.22, examine Figure 4.23 out of a model at 50% DD to see if you can apply those best practices. How many areas are at a high risk for water infiltration?

Scripps Health

Figure 4.23 Roof image

Before we discuss how many areas might lead to water infiltration, let's first look at how many conflicts appear in the model. You'll have to look pretty hard to see one. What you'll notice is that none of the elements in Figure 4.23 are clashing. If you ran clash detection on the duct versus the fire standpipe, you wouldn't have a conflict. They're close, but they're not clashing. Now let's compare that to the number of detail issues (Figure 4.24).

I counted five, but you probably noticed others. Let's take a look at the ones I identified. (Refer to Figure 4.25 for the following list.)

Item 1 The minimum distance from a curb to a wall should be 24″. The duct is only 8″ off the wall and the curb hasn't been modeled. The curb is at least 2″ larger than the duct, as shown in Figure 4.22. If the curb had a cant (triangular shaped block at the bottom of the curb), the distance off the wall could be as little as 0–2″, because you would measure from the bottom of the cant, as shown in Figure 4.21.

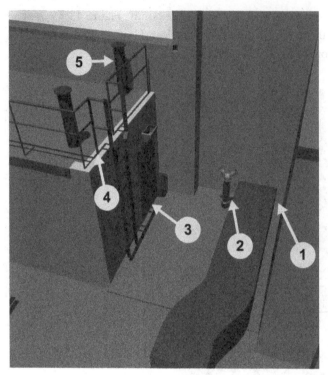

Figure 4.24 Roof image with detail issues

Item 2　The minimum distance the pipe should be penetrating next to a curb is 12″. The fire standpipe is only 10″ away, and again the curb and cant aren't modeled. If they were modeled, the pipe would be anywhere from 2″ to 4″ away from the curb.

Item 3　This is a tricky one, so you get bonus points if you found it. The doorsill is too low for the 8″ inches of overlap required for waterproofing at a vertical face. However, that's only one piece of the issue. Take a look at Figure 4.25, which shows the closest drain to the door.

If the roof doesn't have a natural slope, built-up roofing (commonly referred to as a *cricket*) will be necessary to create a slope to the drain. The drain is approximately 37′ away from the door. At a minimum slope of 1/8″ per foot, there would have to be around 4–5″ inches of built-up roofing at the door before the 8″ of overlap.

Item 4　Four is the "Daily Double," because it wasn't shown in Figure 4.21 or Figure 4.22. The railing system and ladder are both penetrating the flashing that would be required at the edge of the roof. This is another high-risk area for water infiltration that would need to be reviewed with the design team.

Item 5　The pipes that you are seeing are known as *davits*. They are used to support the window-washing system. The davits are only 6″ off the parapet wall. They would need to move off the wall another 6″ or more depending on the cant.

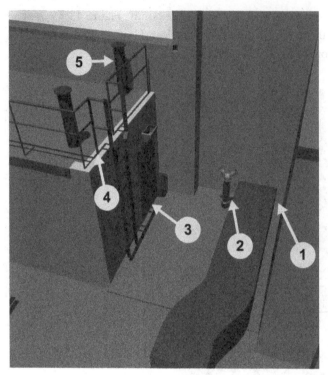

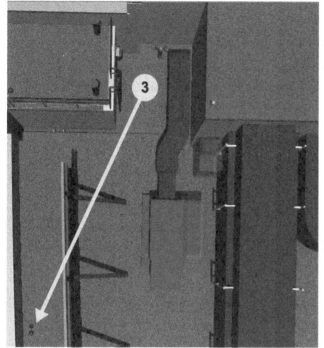

Figure 4.25 Roof drain distance

In the sidebar "The Devil's in the Details," we'll look at the potential impact of item 1 on the design process.

The Devil's in the Details

The steel model has been detailed through the roof, the mill order has been placed, and fabrication is well under way. We're just about to wrap up the Core and Shell package when we spot item 1 from Figure 4.24. To fix the issue, the curb for the duct has to shift over 22" to allow enough room for the roofing material. This causes the fire standpipe to shift 30" to be in compliance with the requirement that there be 12" between the pipe and the curb. We then run clash detection and realize that the duct is now in conflict with the steel, which was previously coordinated around the old curb location.

Now the steel has to be redesigned and refabricated to the new location, and the steel subcontractor will be delivering a change order for the added work and material waste. We also realize that the shaft walls inside the building have to shift out to enclose the new duct location, because the duct is about 16" outside the wall. Because the curb and the shaft wall are shifting, the fire standpipe is now too far out into the stair landing for

Continues

The Devil's in the Details *(Continued)*

code compliance. (Fire standpipes typically route in the stairwells, which are required for egress during emergencies and have a minimum code square footage requirement at landings.) The code has to be met, so now redesign and re-coordination must occur to make sure the stair egress meets code. The butterfly effect creeps back in right when we thought we were going to be able to submit the drawings for permit of the Core and Shell package.

Who knew waterproofing details could have such a big impact on the design? You can imagine what would have happened if that issue hadn't been caught until the construction with RFIs. You'd probably have a disgruntled boss and owner and those items would typically have not have been caught by clash detection. Leveraging the model for this kind of constructability review is invaluable. It is much more efficient than flipping through plans, because everyone's information is consolidated into one location. Using these composite models allows you to get a better understanding of how the systems relate to one another, but you have to leverage the details in order to identify the relationship issues.

Leverage the People

The whole is greater than the sum of its parts—I can't stress that enough. Even if you memorize all the details in *The NRCA Roofing Manual: Membrane Roof Systems*, you'll have only one piece of the complex building puzzle. It may be a corner piece, but it's only one piece. Buildings today are much more complicated than the Empire State Building, so they demand even more planning. Your success is dependent on the construction experience of the entire team and their ability to access and review the model.

Cloud-based BIM software has given confidence to the hesitant and accelerated the learning curve of inexperienced BIM users. In a sense, these solutions almost trick people into using BIM. This factor has had a huge impact on the future of BIM. Up until now, there has been an underlying skepticism from old-school industry professionals toward BIM and the majority of lessons learned are held in the minds of those old-school professionals who claim, "I'm too old to learn that BIM stuff— that's for those techy kids." You've probably heard someone well respected in your organization say something similar. It's a common sentiment among leaders in the construction industry, but I'd be willing to bet those same individuals could take a picture of their kids or grandkids on a smart phone. They may even have a Facebook account where they posted a selfie last week drinking wine in Napa Valley and even used a clever hashtag.

The power of BIM has been simplified down to an elementary level and is so intuitive that a child in Iowa could use an iPad to look at a model of a billion-dollar project being built in Dubai, spin it around, and color on the walls. It truly exemplifies what Thomas Friedman calls the "Flat World," as described in his book *The World Is Flat: A Brief History of the Twenty-first Century* (Farrar, Straus and Giroux, 2005).

One tool that has revolutionized the design process is Autodesk's BIM 360 Glue. This application is a cloud-based collaboration tool. BIM 360 Glue allows you to upload all the team members' models to the cloud (network) and then arrange any combination into a merged (commonly referred to as consolidated or federated) model. All the models are available to every team member through desktop and mobile Glue applications. That means the models are available for review at any time, anywhere, and by anyone. The beauty is that there isn't a long installation process and the majority of people won't realize that they even put Glue on their computer. It works off your Autodesk e-mail account, so if you have old-school individuals on your job, you can set up an Autodesk account for them and invite them to the Glue project through the desktop application, and they'll get an e-mail that looks similar to Figure 4.26.

Reply Reply All Forward

Sat 8/9/2014 11:06 AM

Autodesk BIM 360 Glue <bim360glue-notifications@autodesk.com>

Welcome to Autodesk BIM 360 Glue

To   McCool, David

🛈 If there are problems with how this message is displayed, click here to view it in a web browser.

AUTODESK®
BIM 360™ GLUE

You have been invited to join bim_and_cm_2nd_edition as a host admin.

Get started

Add BIM 360 Glue to your address book to ensure you receive emails. For more information, visit BIM 360 Help.

Figure 4.26 Glue e-mail invitation

Now you can write their Autodesk account information on a sticky note and put it on their computer screen. Once they get the e-mail, tell them to click Get Started and use the sticky note for the login information. Before they know it, they're doing BIM, which they thought was only for those "techy kids." When they see you spin the owner's building around on an iPad, they'll want to create an account so they can do it too.

Once the models have been uploaded, the team can view them, run clash detection, create viewpoints, and mark up constructability issues in real time in the cloud. Whenever someone does any of those items, the whole team is notified (if they want to be) through an e-mail within seconds. Now you can truly leverage the minds of all your team members. This is key to delivering a project that everyone can be proud of.

Uploading a Model to Autodesk BIM 360 Glue

You'll need to download a free trial of BIM 360 Glue from `http://www.autodesk.com/products/bim-360-glue/free-trial`. Once you receive your invitation e-mail, shown in Figure 4.26, follow the instructions to install the desktop application. After you open the application, you'll find a drop-down menu in the top-right corner. Select Admin, shown in the first graphic, and create a new project, as shown in the second graphic.

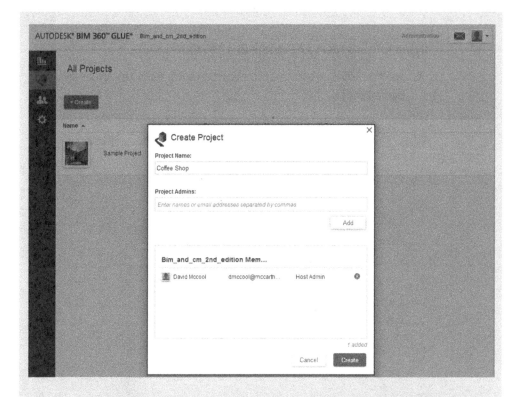

You can import over 50 file formats directly into the Glue application, but it is a best practice to export using the add-ins found at `https://b4.autodesk.com/addins/addins.html`. These add-ins allow you to "Glue" the model out of your authoring software platform. Install Autodesk Revit first and then download the add-in before proceeding with these steps:

1. Download the `Example-50% DD.rvt` file from `http://www.sybex.com/go/bimandcm2e`.

2. Open Autodesk Revit, click Open under Projects, browse to the `Example-50% DD.rvt` file, check Detach From Central, and click Open at the bottom of the window.

3. Click Detach And Preserve Worksets.

Continues

162

Uploading a Model to Autodesk BIM 360 Glue *(Continued)*

4. Select the Addins tab on the top ribbon and click Glue, as shown here, to open the Sign In dialog box:

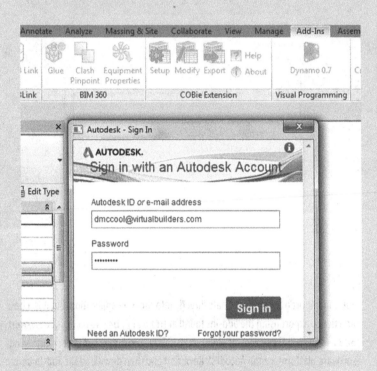

5. Follow the prompts and navigate to the project you created.

6. In the Review And Confirm window, rename the file **Architecture** and create a folder named **50 DD**.

7. Click Glue It.

That's it!

Now you can review the model whenever and wherever you want. Just go to the app store on your mobile device and download the BIM 360 Glue application. Follow the instructions and review the tutorials, and you'll be marking up the model in real time in the cloud before you know it, as shown here:

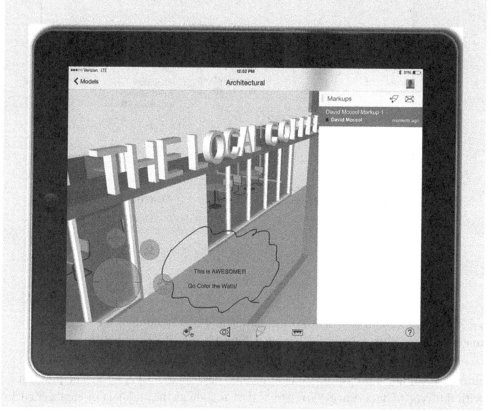

Estimating

Estimating and cost trending with BIM results in tremendous efficiencies for estimators and design managers during preconstruction. Often in the Integrated Project Delivery method, owners will require multiple estimates to be performed at different deliverables to ensure that the project is staying within the originally estimated budget or GMP.

Actual Industry Path of 5D Adoption

The following graphic shows the evolution of model-based estimating, also known by the term *5D*, and our current state.

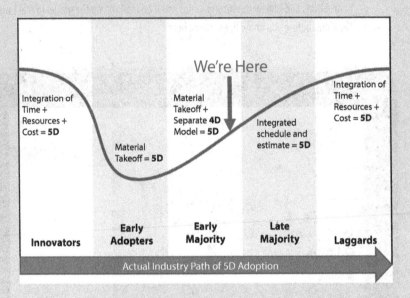

The traditional way of doing estimating and cost trending is for the contractor to get PDFs or printed drawings of each deliverable in order to manually take off (estimate) the work and reach out to the subcontractor community for numbers, similar to bidding a project in the DBB method described in Chapter 2. The problem with this type of data transfer strategy is that as soon as the architect or engineer clicks Print, you are looking at old information. The architect and engineer are not going to stop drawing, tweaking, and modifying the design as the general contractor reviews the documents. The only thing that is current is the model, because it is producing the documents. To truly understand how the design is trending, estimators must leverage the model information.

Revit Schedules for Estimating

Estimating from the model comes down to "garbage in, garbage out." It's pretty straightforward—if the model isn't created accurately, then the quantities will not be accurate, which is where the "I" of BIM becomes important. You must first understand how information is created in the authoring software in order to leverage it for estimating or any other analysis. The model is a database of information that can be manipulated in both a positive and a negative way. The following is a basic example in which you'll create a floor schedule in the Example-50% DD.rvt:

1. Open Example-50% DD.rvt.

2. Type **VG** to open the Visibility/Graphic Overrides window.

3. Deselect all model categories except for Floors and click OK (Figure 4.27).

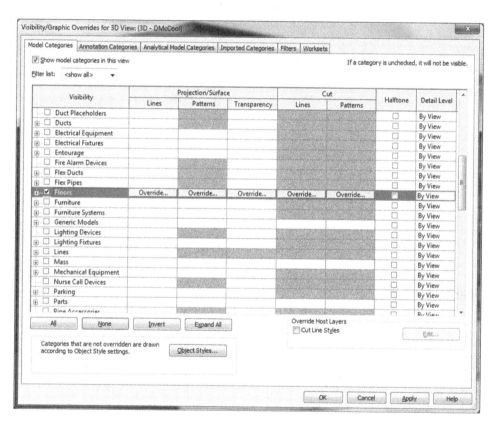

Figure 4.27 In the Visibility/Graphic Overrides window, deselect all categories except for Floors.

4. Select the Manage tab in the top ribbon. Then select Project Units and make sure the format for Volume is cubic yards (CY) and not cubic feet (Figure 4.28).

5. Now select the slab on grade 6″ Concrete Basement Floor and click Edit Type in the Properties window (see Figure 4.29).

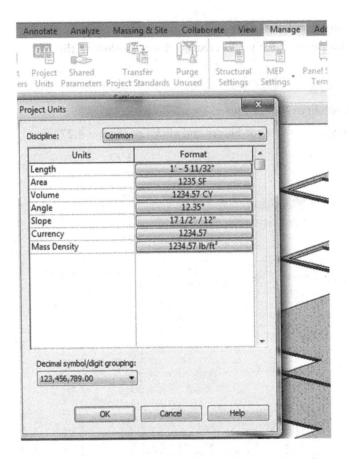

Figure 4.28 Project Units window

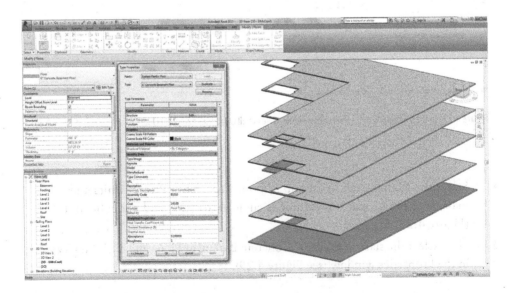

Figure 4.29 Editing the type

In the Type Properties window, you'll start to see the *Information* component of BIM. Notice at the top it tells you the Family (System Family: Floor) and Type (6″ Concrete Basement Floor). Within a Revit family, you have type and instance properties. A type property relates to all the 6″ Concrete Basement Floors for the entire project. An instance property relates to that particular instance of the 6″ Concrete Basement Floor. You can see some of the instance properties on the left-hand side of Figure 4.29. This would include area, volume, thickness, and so forth. Those are unique to the slab you selected, so they're an instance. For estimating, you're typically more interested in the type properties, because you want to take off all of the 6″ Concrete Basement Floors for the entire project.

Look at the Identity Data section in the Type Properties window, and you'll see a Cost field. This is where you'll enter a unit cost for an item. Some companies have cost history data that they use for estimating a project, but I used The Gordian Group's RSMeans to create a unit cost (you can download a free trial at http:// rsmeansonline.com). In RSMeans, I specified 3000-psi ready-mix concrete with a 5% waste factor and used a direct chute for the placement. The cost came out to roughly $140/cy. Once you've filled in the Cost field for the 6″ Concrete Basement Floor, Click OK and repeat the same process for the elevated slabs. You only have to do one, because you're adjusting the type property and not the instance property.

6. After you've entered the unit costs for both types of floors, select the View tab on the ribbon. Click the Schedules menu and select Schedule/Quantities to open the New Schedule dialog box (Figure 4.30).

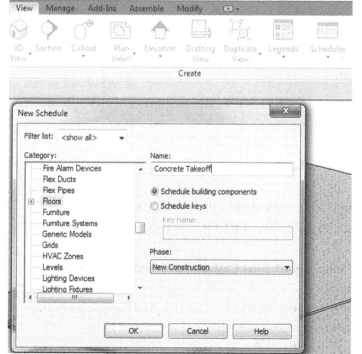

Figure 4.30 The New Schedule dialog box

7. Click Floors, and enter **Concrete Takeoff** in the Name text box; then click OK.

8. In the Schedule Properties window, scroll through the Available Fields list box and add the parameters Area, Cost, Family And Type, and Volume.

9. Click Calculated Value to open the Calculated Value dialog box. Here you can create formulas with the added properties.

10. Enter the name **Concrete and Placement**.

11. In the Formula field, click the ellipsis to open the existing properties.

12. Create the formula shown in Figure 4.31.

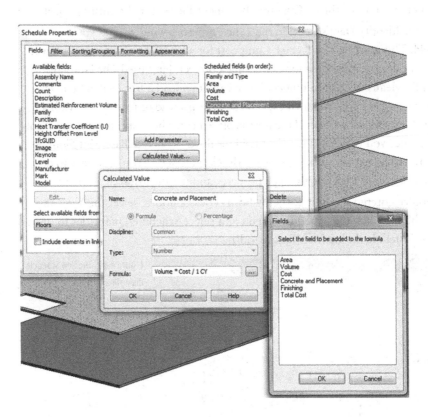

Figure 4.31 Creating the Concrete and Placement formula

13. Repeat steps 9-11 and create a calculated value named **Finishing**.

   For Finishing, use the formula **Area*.44/1sf** (I used RSMeans to obtain the finishing unit cost).

14. Create a calculated value named **Total Cost** with a formula **Concrete and Placement+Finishing**.

15. In the Schedule Properties window, select the Sorting And Grouping tab and click the Grand Totals option (Figure 4.32).

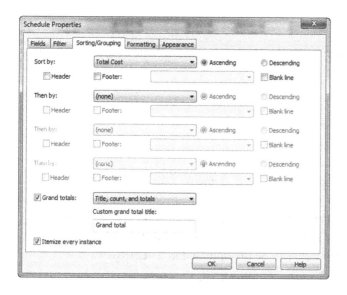

Figure 4.32 Select the Grand Totals option.

16. Click the Formatting tab and select the Concrete And Placement field.

17. Select the Calculate Totals option.

18. Click the Field Format button. Enter the settings shown in Figure 4.33 and click OK.

19. Repeat steps 16-18 for both Finishing and Total Cost.

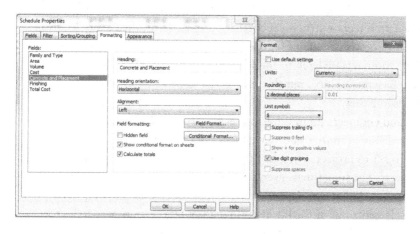

Figure 4.33 Enter the field formatting settings shown here.

20. Once you've completed the formulas, click OK to generate the schedule, shown in Figure 4.34.

<Concrete Takeoff>						
A	B	C	D	E	F	G
Family and Type	Area	Volume	Cost	Concrete and Placement	Finishing	Total Cost
Floor: 6" Concrete Basement Floor	6873 SF	127.27 CY	140.00	$17,817.70	$3,023.92	$20,841.62
Floor: 1-1/2" Mtl Deck w 3" Concrete	6594 SF	91.59 CY	148.00	$13,555.22	$2,901.55	$16,456.77
Floor: 1-1/2" Mtl Deck w 3" Concrete	6594 SF	91.59 CY	148.00	$13,555.22	$2,901.55	$16,456.77
Floor: 1-1/2" Mtl Deck w 3" Concrete	6594 SF	91.59 CY	148.00	$13,555.22	$2,901.55	$16,456.77
Floor: 1-1/2" Mtl Deck w 3" Concrete	6594 SF	91.59 CY	148.00	$13,555.22	$2,901.55	$16,456.77
Grand total: 5				$72,038.58	$14,630.12	$86,668.70

Figure 4.34 The Concrete Takeoff schedule

You can follow the same process for every element in the model to create a total project cost. The first time you create a schedule, the process may seem a bit cumbersome, but once you've done it a couple of times, you'll find it easy. When you leverage the data in a positive way, the results can be very powerful and efficient.

Now let's look at how the data can be manipulated in a negative way:

1. Go back to the 3D view by clicking on the house icon at the very top left of the screen or minimizing the schedule view.

2. Select the slab on grade (6″ Concrete Basement Floor) and select Edit Type in the Properties window (Figure 4.29).

   Note the instance properties.

3. Click the Rename button.

4. Change the name to **12″ Concrete Basement Floor.**

5. Click OK.

You'll notice that the volume and thickness didn't change but the type name has changed. This name (data) is tied to every schedule and tag in the model. What that means is that if the architect or structural engineer clicked Print, then every location where this basement floor is called out on the drawings would say 12″ Concrete Basement Floor. If you navigate back to the schedule, you'll see what I mean. It now says 12″ Concrete Basement Floor, but the price is the same as the 6″ floor. That's an almost $18,000 discrepancy.

Architects and engineers do multiple projects so they will develop robust libraries of standard Revit families in order to help their teams be more efficient and consistent from project to project. On a fast-track design schedule, these standard libraries are lifesavers, but sometimes, even with the libraries, designers use shortcuts to meet the deliverables. What can happen is that experienced and inexperienced modelers will use elements out of the library and manipulate certain properties to fulfill the design intent and neglect properties required for estimating, like the structure of an element. You can see these properties by clicking the Edit button in the Structure field shown in Figure 4.29. Notice how it says 6″ under the Thickness column even though the type says 12″ Concrete Basement Floor. There's a conflict within the model information, but the *design intent* is met for the drawings (Figure 4.35).

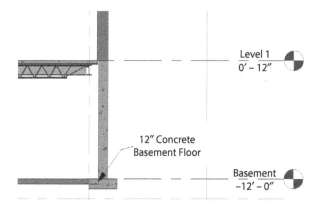

Figure 4.35 Elevation view

Hopefully estimators would catch this in the manual takeoff, but they may just look at the tag (12″ Concrete Basement Floor) and do a square foot calculation. This is part of the reason why Architect and Engineering firms are hesitant to release their models to contractors. First, they're giving you their company's library of families, and second, they don't want to be liable for errors in information that were extracted outside the printed contract drawings.

On the August 8, 2014 "Fridays with Trimble" webinar, Vico Office product manager Duane Gleason said that on average you can extract about 65 percent of the estimate from an architect or engineer's model. To mitigate the 35 percent risk, some companies create their own parallel model as they receive information from the architect and engineer to ensure the model is accurate for estimating. This isn't necessarily the leanest approach, since you're doubling the efforts of model creation, but it will help identify constructability issues and the estimate will be more reliable. Others prefer to use a hybrid approach of 2D and 3D estimating, similar to constructability review. This seems to be the most popular approach, and both Autodesk and Trimble have developed software programs for this takeoff method.

Cost Trending with Assemble

Cost trending during design is a different strategy than estimating precise quantities. Cost trending is more about being able to track the evolution of the design as a rough order of magnitude (ROM). Basically, you acknowledge that the model isn't perfect and leverage it for its efficiencies, such as the ability to produce instant quantities. There are a couple ways to approach this idea:

Sharing Your Company's Pricing One way is to have a collaborative team where you share your company's unit pricing with the design team so they can create calculated values within the evolving models. This allows the entire team to see real-time gains and losses. The design team can also export the schedules as text files that can be copied into Microsoft Excel or any other database for further analysis (Figure 4.36).

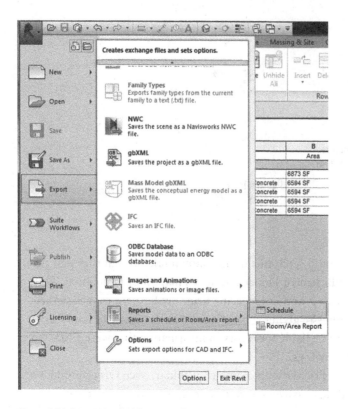

Figure 4.36 Exporting a schedule to a text file

However, in order for this to be successful, there has to be buy-in from the design team to create calculated values for all their elements, which might be more than they bargained for in their base fee on a Design-Assist project. It also requires you to share your company's cost histories, which are sometimes held "close to the vest," because you don't want to let the competition know what cards you're holding. This requires trust with the design team and can limit the ability for this method to be used on a Design-Build or IPD.

Creating Your Own Schedules The other option is to create the schedules yourself. This approach can seem daunting, especially for estimators new to BIM. It can also be inefficient to repeat the data entry every time you receive a new Revit model from the design team.

Fortunately, cloud-based solutions have done it again and are turning old-school estimators into "new-school BIMstimators." In the next example, I'll be using Assemble to demonstrate how to leverage cloud-based tools for cost trending through the design process. Assemble is a web-based data management tool that automates the scheduling process and allows you to manipulate the data in the cloud for estimating or model management. You'll need to download a trial of Assemble from www.assemblesystems.com/downloads to install the add-in. Make sure Revit is closed when you do.

1. In Revit, open Example-50% DD.rvt.

2. Select the Assemble tab in the top ribbon.

3. Select the Publish tab (it has an arrow pointing up) to post the model to the cloud. Then, click the View button.

   Notice how the interface looks very similar to the schedule that you created in Revit (Figure 4.37). The Name column is the Family Name and Type column that was in your Revit schedule. Assemble just extracted the data and created a schedule of all the model elements.

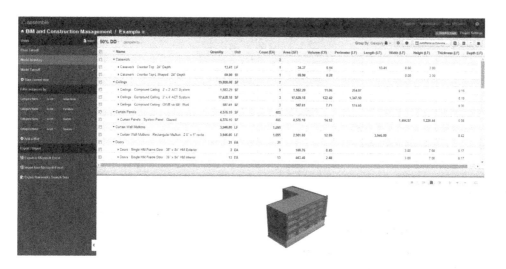

Figure 4.37 Assemble interface

4. Click Add/Remove Columns in the top-right corner and remove the columns Perimeter, Width, Height, Thickness, and Depth.

5. Add columns Unit Cost and Total Cost. You can use the search field. Then click Apply.

6. Click the (-) symbol in the top right to collapse the families.

7. Click through the families to see how the model is linked to the rows.

8. Expand the Doors category and select 36 × 84 HM Exterior.

9. On the right side of the screen is an information panel with an Instance and a Type tab. Under the Instance tab you'll see the Unit Cost field. Enter a ROM of $2100 and then click Save.

10. Use RSMeans or plug in your own numbers for the rest of the model elements. The estimate should look similar to Figure 4.38.

50% DD ▾ compare to   Group By: Category

Name	Quantity	Unit	Count (EA)	Area (SF)	Volume (CY)	Length (LF)	Unit Cost	Total Cost
▸ Curtain Panels : System Panel : Glazed	4,576.10	SF	405	4,576.10	14.12		90.00	411,849.08
▾ Curtain Wall Mullions	3,946.80	LF	1,095					
▸ Curtain Wall Mullions : Rectangular Mullion : 2.5" x ...	3,946.80	LF	1,095	2,561.80	12.69	3,946.80		
▾ Doors	31	EA	31				▪	78,400.00
▸ Doors : Single HM Frame Door : 36" x 84" HM Exte...	3	EA	3	100.76	0.45		2,100.00	6,300.00
▸ Doors : Single HM Frame Door : 36" x 84" HM Interior	13	EA	13	443.40	2.48		1,800.00	23,400.00
▸ Doors : Single HM Frame Door : 36" x 96" HM Exte...	8	EA	8	318.01	1.77		2,150.00	17,200.00
▸ Doors : curtain_wall_single_glass : 10" Bottom Rail	7	EA	7	194.10	0.58		4,500.00	31,500.00
▾ Floors	493.63	CY	5				▪	191,087.18
▸ Floors : Floor : 1-1/2" Mtl Deck w 3" Concrete	366.36	CY	4	26,377.73	366.36		400.00	146,542.93
▸ Floors : Floor : 6" Concrete Basement Floor	127.27	CY	1	6,872.54	127.27		350.00	44,544.25
▸ Generic Models	9	EA	9					
▾ Plumbing Fixtures	24	EA	24				▪	73,200.00
▸ Plumbing Fixtures : Toilet-Domestic-3D : Toilet-Do...	12	EA	12	243.42	0.49		3,500.00	42,000.00
▸ Plumbing Fixtures : Urinal-Wall-3D : Urinal-Wall-3D	4	EA	4	20.92	0.05		2,800.00	11,200.00

6 Selected Instances

Figure 4.38 Complete ROM estimate

11. Close Assemble.

12. Open Example-75% DD.rvt.

13. Click the Publish tab to export to Assemble.

14. Once you publish the 75% version, Assemble asks if you want to view it. Click Compare To and select the 50% DD.

15. Add the columns Unit Cost and Total Cost back in.

    Notice how Assemble kept the unit cost and shows you the deltas between the two increments. There were significant changes to the exterior skin systems.

16. Repeat steps 12–14 to compare the Example-75% cd.rvt to the 75% DD. You should see results similar to those in Figure 4.39.

What you'll notice is that from 50% DD to 75% DD you went up in some areas but brought the pricing back down a little in the 75% CD. Hopefully your GMP was calculated off 75% DD and not 50% DD or you'll have quite a bit of value analysis to do in order to bring your number back where it needs to be.

Here's your workflow: Designer or BIM manager exports the model to Assemble. Estimator then logs into Assemble.

▼ Name	Quantity				Unit Cost			Total Cost		
	75% DD	75% CD	Variance	Unit	75% DD	75% CD	Variance	75% DD	75% CD	Variance
▸ Ceilings	17,628.18	19,589.70	(1,961.52)	SF	3.50	3.50	0.00	61,698.62	68,563.95	(6,865.32)
▸ Curtain Panels	567.11	259.52	307.58	SF	90.00	90.00	0.00	51,039.81	23,357.11	27,682.50
▸ Curtain Wall Mullions	196.00	98.00	98.00	LF			0.00			0.00
▸ Floors	493.63	593.06	(99.43)	CY			0.00	191,067.18	229,612.64	(38,525.46)
▸ Railings	456.82	447.66	9.16	LF	175.00	175.00	0.00	79,943.36	78,339.93	1,603.45
▸ Roofs	6,805.28	8,143.61	(1,338.33)	SF	13.50	13.50	0.00	91,871.25	109,938.75	(18,067.50)
▸ Structural Columns	29.73	25.68	4.05	TON	4,000.00	4,000.00	0.00	118,925.01	102,707.96	16,217.05
▸ Structural Foundations	22.06	28.17	(6.11)	CY	350.00	350.00	0.00	7,721.06	9,859.95	(2,138.89)
▸ Structural Framing	1.02	1.88	(0.87)	TON	4,000.00	4,000.00	0.00	4,071.31	7,534.80	(3,463.49)
▸ Walls					▪	▪	0.00	1,163,158.91	1,161,064.08	2,094.82

Figure 4.39 Cost trending 75% DD to 75% CD

Voilà, he just became a "BIMstimator" with a username and password. Now your estimator doesn't have to wait until the designers click Print. They can be "trending" the model every day, week, or month without downloading any BIM software to their computer. Again, this approach recognizes the flaws in the model but leverages it so that you can determine the red flags and dig deeper into those issues. It creates transparency between the contractor, designers, and owner. Everyone can see why the design is trending up or down.

> **Note:** Assemble is a quick and useful tool, but there are other features that we will not be able to discuss in this book. Go to http://assemblesystems.com to learn more about this tool.

Analysis

The last use we'll talk about in preconstruction is analysis. This is a very broad term because of the amount of information that can be used from a model. For instance, Autodesk 360 Glue and Assemble could technically be considered analysis software, because they're extracting and analyzing data created in the authoring software. In fact, any software program outside of the authoring platform that is using its information could be termed analysis software. Analysis could simply mean showing the owner aesthetics of the exterior in 3D, or it could be more complex, like running an acoustic study in an auditorium to see how sound reverberates in the room or creating a pedestrian traffic study to see how people flow in a busy airport terminal. To bring order to this term, people started clustering analysis software into various dimensions of BIM, such as 3D (visual/spatial), 4D (schedule), 5D (cost), and 6D (facilities management). Autodesk 360 Glue would be considered 3D, Assemble would be considered 5D, and so on. However, certain programs fall outside these dimensions. Some have coined them as "7D," taking them to another dimension. It seems a little too sci-fi for me, so we'll refer to them as the true "analysis" tools. These are the programs that study sustainability and the building's impact to our planet and people.

The 2030 Challenge

Our planet, building industry, and human behavior were forever changed by the popularity and use of artificial thermal comfort and lighting starting around the 1930s. Designers had freedom to explore materials from around the world and to create buildings that didn't have to rely on passive heating/cooling or natural lighting. We could stretch the limits and build anything we wanted to anywhere in the world. Air conditioning (A/C) was no longer exclusive to the social elite, and the Census Bureau household survey in 1960 showed that 6.5 million A/C units were in households and the market was just starting to take off. The Roosevelt administration established the Rural Electrification Administration (REA) in 1935 and by 1960 "96 percent of American farms were connected to electric lines," according to Jane Brox in *Brilliant: The Evolution of Artificial Light* (Mariner Books, reprint edition, 2011). The working class was more productive and could work all hours of the day without being hindered by bad weather or lack of daylight. It's no wonder that Richard L. Evans coined the term "workaholic" in the 1960s. The results of this electrical dependency shouldn't come as a shock either.

According to the U.S. Department of Energy's *2010 Buildings Energy Data Book*, the building industry in the United States used up about half of the nation's total energy due to various building end uses. Nearly 70 percent of our energy demands are the result of heating, cooling, water heating, and lighting (shown in Figure 4.40).

The organization Architecture 2030, a nonprofit, nonpartisan, and independent organization, was founded by Ed Mazria in 2002. In a Greenbuild 2013 Master Series Session, Mazria showed statistics from The McKinsey Global Institute projecting that we will build and renovate approximately 900 billion square feet of buildings by 2030, which creates an opportunity to reshape the built environment and our dependency on fossil fuels for energy (http://vimeo.com/81627798).

The 2030 Challenge was established by Architecture 2030 to create a roadmap to achieving net zero buildings by 2030. The 2030 Challenge mandates that every new or renovated building "shall be designed to meet a fossil fuel, GHG-emitting, energy consumption performance standard of 70% below the regional (or country) average/ median for that building type" from 2015 to 2020. From 2020 to 2025, the buildings must be 80 percent below the average; from 2025 to 2030, 90 percent; and from 2030 and beyond, net zero. They believe if we can accomplish these goals we can reduce the dependency of fossil fuels and save our planet for future generations.

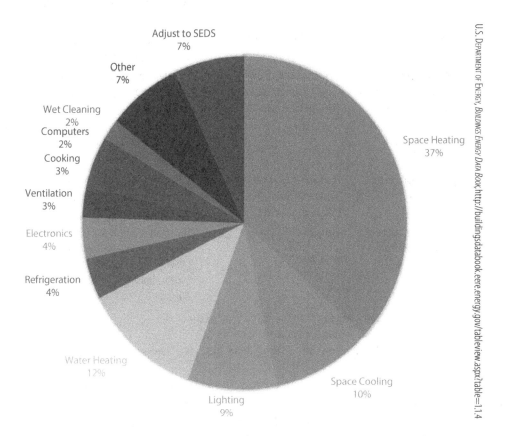

U.S. DEPARTMENT OF ENERGY, *BUILDINGS ENERGY DATA BOOK,* http://buildingsdatabook.eere.energy.gov/tableview.aspx?table=1.1.4

Figure 4.40   2010 U.S. buildings energy end-use splits by fuel type

Overview of Sustainability and BIM

The 2030 Challenge has obviously become a popular topic among the AEC industry and governmental agencies, but it only represents one aspect of sustainability. The term "sustainability" encapsulates all aspects of the qualitative life-cycle cost of a building. At a minimum this includes site selection, water use, energy use, material use, and the quality of life for the tenants. Is the site existing or are you disturbing a natural habitat to build? Are you collecting rainwater for irrigation? Will the offices have natural light? Where are the materials coming from? How are the materials manufactured? These are the kinds of questions that must be analyzed to create a high-performance sustainable building, and a number of building rating systems around the world address these concerns. Table 4.2 shows some popular ones.

Organization	Rating system	Criteria
United States Green Building Council (USGBC)	Leadership in Energy and Environmental Design (LEED)	Site Selection, Water Use, Energy, Indoor Quality, Materials and Resources, Commissioning
International Living Future Institute	Living Building Challenge	Site Selection, Water Use, Energy, Health Materials, Equity, Beauty
Building Research Establishment (BRE) Group	Building Research Establishment Environmental Assessment Method (BREEAM)	Management, Health, Energy, Transport, Water Use, Materials, Waste, Land Use, Pollution
Green Building Initiative	Green Globes	Project Management, Site, Energy, Water, Materials and Resources, Emissions, Indoor Environment
Japan GreenBuild Council	Comprehensive Assessment System for Built Environment Efficiencies (CASBEE)	Building Environmental Efficiency = Built Environment Quality/Built Environment Load (BEE=Q/L)
International Initiative for a Sustainable Built Environment (iiSBE)	SBTool	Site Location, Site Regeneration and Development, Energy, Environmental Loadings, Indoor Quality, Service Quality, Social/Cultural/Perceptual Aspects, Cost Aspects

www.usgbc.org/leed, http://living-future.org/lbc, www.breeam.org, www.greenglobes.com/home.asp, www.ibec.or.jp/CASBEE/english/, www.iisbe.org/sbmethod

Building rating systems have changed the construction industry's outlook on sustainability. The growth rate of the USGBC and LEED is a testament to that. The USGBC started in 1993 and has now grown into the world's most widely recognized rating system. There are now 76 chapters, 12,800 member organizations, and almost 200,000 LEED professionals, according to the 2012 statistics. "LEED is certifying 1.5 million square feet of building space each day in 135 countries" (www.usgbc.org/about). So with all this momentum, how are we still warming the planet?

Well, there are a couple of loopholes to the rating systems; they're voluntary and you can select which credits to go after. The question owners start asking is, "How do I get to LEED Gold for the lowest cost?" as opposed to, "What are the best ways to reduce my building's CO_2 emissions?" Owners realize that they can charge higher rent for a LEED-certified building—they have happier tenants and a better resale value. It's business. For instance, I installed bike racks, locker rooms, and showers in a data center in the middle of Alabama. You'd have to be preparing for the Tour de France to ride a bike out there, but it was inexpensive and gave us points toward LEED Gold. It is for this reason that new and modified building codes have been created to enforce sustainability as a requirement, as shown in Table 4.3.

Code	Requirements
IgCC—International Green Construction Code	Site Selection, Water Use, Energy, Indoor Quality, Materials and Resources, Commissioning
Title 24—California Energy Commission	Building Envelope, HVAC, Indoor/Outdoor/Sign Lighting, Power, Solar Ready, Covered Process (Air Quality), Performance Approach, Commissioning
ASHRAE 189.1—Standard for High Performance Green Building	Site Selection, Water Use, Energy, Indoor Quality, Materials and Resources, Commissioning
Part L (L1A)—Conservation of Fuel and Power in New Dwellings	HVAC, Lighting, Building Fabric
ASHRAE 90.1—Energy Standard for Buildings	Exterior Skin, HVAC, Service Water Heating, Power, Lighting, and Other
ASHRAE 62.1—Indoor Air Quality (IAQ) Regulations	Air Flow and Indoor/Outdoor Air Ratios
ASHRAE 55—Thermal Conditions for Human Occupancy	Temperature and Humidity of Indoor Spaces

```
www.iccsafe.org/CS/IGCC/Pages/default.aspx, www.energy.ca.gov/title24/,
www.planningportal.gov.uk/buildingregulations/approveddocuments/partl,
https://www.ashrae.org/standards-research--technology/standards--guidelines
```

These codes serve two purposes: first, governmental authorities and owners can demand minimum sustainability requirements, and second, the rating systems can continue to create scoring based on bettering these code requirements. The code works in tandem with the building rating systems, which is why the IgCC and ASHRAE 189.1 were developed with aid from the USGBC, among others. The establishment of these codes elevates the necessity for our industry to change, because we're not changing fast enough.

BIM can be used in conjunction with these rating systems and codes to analyze designs. You can leverage the information using the same principles as estimating to get basic sustainable calculations.

Calculating Concrete CO_2 Emissions Using Revit

Concrete, in its simplest form, is made from Portland cement, aggregates, and water. Cement manufacturing began to contribute to greenhouse gas (GHG) calculations in the late 1940s. The reason for this is because the heat required to produce one ton of Portland cement creates one ton of carbon dioxide (CO_2). This is why you'll see rating systems give credits for fly ash content in concrete. Fly ash is a byproduct of burning coal and can be

Continues

Calculating Concrete CO_2 Emissions Using Revit *(Continued)*

used to offset the amount of cement in concrete (LEED counts fly ash towards recycled content in *Materials and Resources*).

Practice what you've learned!

Let's assume your concrete mixture weighs 3,500 pounds per cubic yard. Cement makes up about 16 percent of the overall weight of concrete and the CO_2 emissions from cement are a 1:1 weight ratio. Use the volume of concrete and two calculated values to estimate the total tonnage of CO_2 emissions for the 50% DD model concrete design. It should look similar to the following illustration.

<50% DD Concrete Emissions>			
A	**B**	**C**	**D**
Family and Type	Volume	Cement Weight	Tons of CO2
Floor: 6" Concrete Basement Floor	127.27 CY	71,271	36
Floor: 1-1/2" Mtl Deck w 3" Concrete	91.59 CY	51,290	26
Floor: 1-1/2" Mtl Deck w 3" Concrete	91.59 CY	51,290	26
Floor: 1-1/2" Mtl Deck w 3" Concrete	91.59 CY	51,290	26
Floor: 1-1/2" Mtl Deck w 3" Concrete	91.59 CY	51,290	26
Grand total: 5			138

Schedules in Revit can be used for calculating various credits related to materials (recycled content, reuse percentage, regional location), water use (fixture flow rates, area of rainwater collection), and site use (heat island effect, stormwater design), but the more complex analysis for energy and quality of life requires software or add-ins with greater capabilities. This gets into the thermal and visual comfort and is the crux of our global crisis. It is also where BIM can have the most impact.

According to the Office of Energy Efficiency & Renewable Energy (EERE) website (http://apps1.eere.energy.gov/buildings/tools_directory/), there are as of this writing 417 different kinds of analysis programs on the market for energy and sustainability studies. You'll find the list of programs and their uses at the EERE site. We explored these tools in Chapter 2, and their use demands a high level of understanding of the interdependencies between systems (mechanical, electrical, and architectural). This analysis must occur during the conceptual design in order to achieve a truly sustainable building.

A few items have to be determined prior to starting analysis models (shown in Figure 4.41). These items include Location, Owner Goals, Owner Program, Site Constraints, Code Requirements, and Type of Building. These decisions will steer the sustainability design options and strategies. If you're building in a city where sunlight is being blocked by other buildings but the owner wants tenants to have natural light, you might have to get creative with larger windows, light shelves, skylights, or solar tubes to bring natural light into the building for the tenants. In contrast, if you're building in the desert you'll want to optimize your glazing and use more shading strategies to minimize heat gain. Checking out these six categories prior to orientation and massing studies will help narrow down the design options.

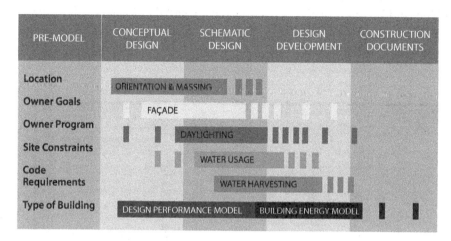

Figure 4.41 Sustainability analysis schedule

The book *Green BIM* by Eddy Krygiel and Bradley Nies (Wiley, 2008) is dedicated specifically to BIM, green construction, and design strategies and is an excellent resource if you want to dig more in depth into sustainability strategies and how you can leverage BIM during the design. This book focuses heavily on the importance of orientation, massing, and glazing. These three factors are the foundations for the success of a sustainable building and reducing GHG emissions. "The most powerful of Mother Nature's resources is the sun. The sun provides us three key resources; light, heat, and power," according to the authors. It's during the conceptual and schematic design that you want to maximize the use of this resource to minimize the use of fossil fuels.

At the end of *Green BIM*, the authors talk about the future of BIM and the opportunity it will provide in restoring our planet:

> *Parametric modeling will go well beyond mapping relationships between objects and assemblies. Both model and designer will have knowledge of climate and region. The model will know its building type, insulation values, solar heat gain coefficients, and impact on the socioeconomic environment it resides within. It will inform the design team with regard to upstream impacts and downstream consequences of their choices. As the building is modeled, prompts will inform the designer of the impact of the building orientation and envelope choices on the sizing of mechanical system and the comfort of its inhabitants.*

Eddy Krygiel and Bradley Nies, *Green BIM* (Wiley, 2008)

That was written in 2008 and guess what? Thanks to big strides in new BIM software, we're now able to do that kind of analysis.

Sustainability Analysis with Sefaira

Sefaira is a plug-in for Revit and Trimble SketchUp that allows for real-time energy and daylighting feedback as you design. When you open the program, you can select the type of building and location. Sefaira will automatically pull the closest weather data information for calculations. You can see in Figure 4.42 that the building is going to be an office located in New York, New York.

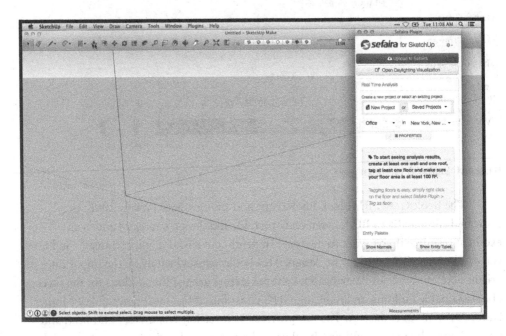

Figure 4.42 Sefaira interface

As you begin to model, Sefaira will run algorithms behind the scenes to interpret the model, predicting what will be the floor, wall, and roof. The model in Figure 4.43 is a simple shoebox design of a 10-story building. Once content is generated, the analysis will automatically begin. Because the building is in New York without windows, the analysis is telling us that there's a significant heating load and the building is underlit, which is shown in the Real Time Analysis window in Figure 4.43.

Let's add some glazing on the south façade to see what it will do to our loads. To add glazing, simply color any face with a translucent material. The materials can be found by clicking on the icon that looks like a paint bucket. Sefaira uses that information to predict the surfaces you want to add glazing to, as shown in Figure 4.44. If at any point the interpretation of the surface is wrong, you can override it using the Entity palette to change the properties.

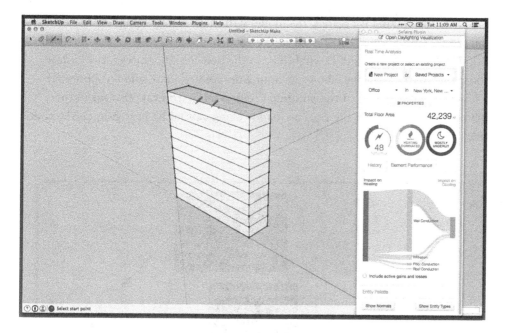

Figure 4.43 Shoebox design without glazing

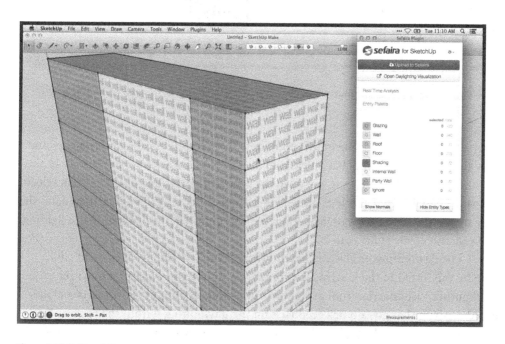

Figure 4.44 Entity palette

Once the glazing is created, the model will immediately update with new calculations. You'll notice three dials in the Sefaira window: Energy Use Intensity (EUI), Energy Segments, and Daylighting. In the new design strategy, the building is under the 2030 Challenge, shows a cooling dominated load, and has potential glare issue in some spaces of the building (Figure 4.45). You can also see in the graph below the dials that the cooling load is being caused by heat gain on the south façade.

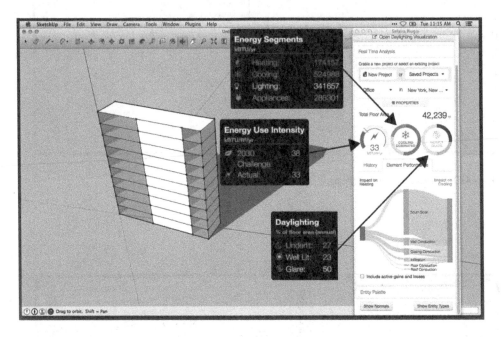

Figure 4.45 Energy and daylighting analysis

You may be wondering at this point how the software is calculating energy and daylighting. Sefaira is using baselines from the codes that were discussed earlier in this section (shown in Table 4.3). If you open the properties drop-down list, you can select preloaded baselines to run the analysis (Figure 4.46). Picking the baseline populates the code minimum values for wall insulation, floor, roof, mechanical systems, and lighting. You can also create your own.

The daylighting analysis is based off the LEED v4 requirements and can even be submitted to the USGBC for credits. To do so, you must first open the Daylighting Visualization tool at the top of the Sefaira plug-in. This starts a cloud-based analysis that gives quick feedback on the daylighting within the spaces. You can look at the daylighting based on annual availability or time of day (Figure 4.47 and Figure 4.48).

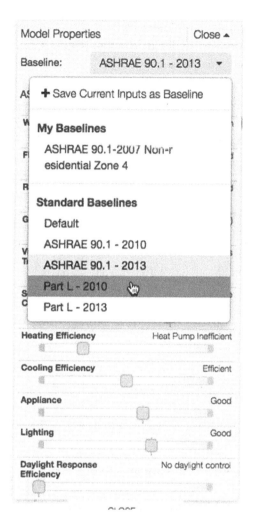

Figure 4.46 Building code selection

This program allows designers to stay in their comfort zone while giving them instant feedback to make good design decisions. It also allows designers to quickly test different orientations, heights, glazing ratios, and shading strategies in order to find the optimal design. Once they have a design they like and they are sure it meets both aesthetic and performance requirements, they can upload the model to the Sefaira web application for further detailed analysis (Figure 4.49). It's at this point that a more experienced sustainability consultant may be required.

The web application runs off a combination of the Sefaira and U.S. Department of Energy's EnergyPlus calculation engine. It analyzes the initial information from the SketchUp model, but it allows you to manipulate that information to run more detailed analysis strategies for Renewables, Water Fixtures, Envelope, and HVAC (shown on the left-hand side of Figure 4.49). You can also bundle the strategies to see the collective impacts on the heating/cooling loads, energy consumption, CO_2 emissions, and annual utility bills. The program has a compare function, similar to Assemble, that allows you to analyze various design options or trend the performance of design iterations (Figure 4.50). And last but not least, you can run water efficiency analysis studies.

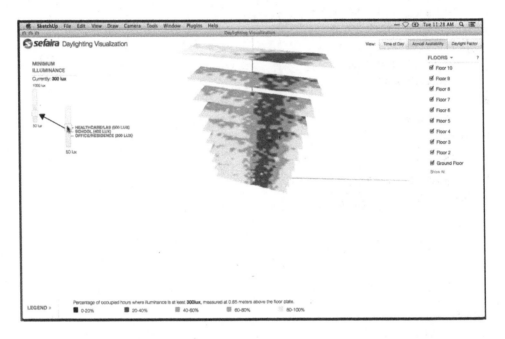

Figure 4.47 Daylighting Annual Availability

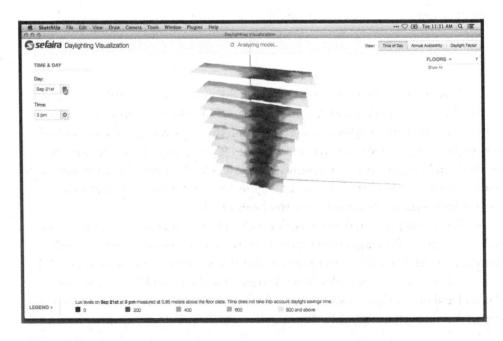

Figure 4.48 Daylighting Time of Day

Figure 4.49 Sefaira web application

Figure 4.50 Design option comparison

In a phone conversation with Carl Sterner, Senior Product Marketing Manager, we discussed the limitation of Sefaira. At this point the program has robust features for building envelope studies, but it is lacking in full energy analysis. However, Sefaira is currently in the process of developing Sefaira for Engineers, which according to Sterner is "intended to provide early-stage analysis of mechanical equipment sizing, space requirements, and energy performance for different system options, allowing engineers to collaborate more effectively with design teams from the earliest stages of a project." This product is expected to launch in 2015.

Logistics and Planning

Site logistics plans begin at the pursuit phase and carry through the construction phase (Figures 4.51–4.54). These plans are especially important when you are dealing with dense urban environments or challenging sites, as discussed in Chapter 3. BIM offers tools to develop plans for erosion control mitigation, crane logistics, material staging areas, vehicular traffic/access, material hoists, equipment, scaffolding, and safety. A number of software programs can produce these plans, but probably the most popular and efficient ones are Trimble SketchUp and Autodesk InfraWorks 360. The following figures are examples of site logistics plans produced from these programs.

Figure 4.51 Site logistics plan using Trimble SketchUp (rendered with Autodesk 3ds Max)

Figure 4.52 Site logistics plan using Trimble SketchUp

Figure 4.53 Site logistics plan rendering using Trimble SketchUp (rendered with Autodesk 3ds Max)

COURTESY OF McCARTHY BUILDING COMPANIES, INC.

Figure 4.54 Site logistics plan rendering using Autodesk InfraWorks 360

Summary

This chapter has been a journey from past to present on how to approach preconstruction. The workflows demonstrated should provide foundational knowledge for leveraging BIM during the design phase. BIM's ability to inform the design and construction team on early decisions will make the preconstruction phase continue to grow in importance as the project moves toward construction.

The goal of using BIM in preconstruction should be to make the construction phase boring. When you think about it, that's what BIM has the potential to do: automate construction. By analyzing a structure down to the elemental level before a shovel hits the ground, we can ask more questions and find more answers. Tools like Sefaira and DSM will become the new "business as usual" and provide us with buildings and structures that perform better. The ability to analyze and coordinate construction models is leading to tremendous gains in construction efficiencies. The next chapter will discuss these gains and the various ways the model information is being leveraged during construction.

BIM and Construction

5

BIM is a powerful tool during construction. Used in conjunction with a digital documentation workflow, BIM in the construction phase of a project gives the project staff a better way of working and provides many tools that were not available before.

Overview of BIM in Construction

BIM in construction has changed dramatically over the last five years. This is in large part accredited to a renewed focus and energy in the construction market space. Having achieved a relatively high volume of adoption in BIM across the design landscape, software vendors began to focus on the next target market for BIM tools- construction firms. Although this was probably the logical sequence of product deployment because of immediate value, in some ways the order of deployment from these more design-focused tools to construction was a "beat to fit" solution. The industry is currently reversing some of these initial assumptions that included the belief that the tools designers used would also be the same tools required of construction teams. Now the software industry is looking specifically into how BIM applications can create value as a tool for construction to achieve better results.

The BIM software industry continues to develop modeling tools. These tools are providing contractors with ways of finding and coordinating construction as well as identifying gaps in information exchange from model platforms to other systems such as estimating, scheduling, etc. In recent years, large software vendors have focused on the development of tools for construction firms specifically. This chapter provides insight into how these available BIM tools can be used and offers suggestions about the areas in which software still needs to develop to bridge the gap between design and construction. It should be stated that BIM in fact does "work" in the field. It is a myth that BIM isn't for the field. There is plenty of room for improvement and smoother interoperability between field systems and tools, but a great opportunity is overlooked by contractors who choose not to use today's technologies and leverage them to some extent on the construction site.

The model coordination plan continues to play a critical role in a BIM process; what was developed at the beginning of the project defines who uses the model, where the model gets distributed, and what it's used for during construction. In the field, BIM is used to:

- Analyze physical construction information
- Manage in-field clash detection
- Update model-derived estimates (5D)
- Clarify scope and work packages
- Manage material inventory
- Perform 4D scheduling updates
- Create a field sequencing clash detection
- Clarify the installation of fabricated components
- Enhance on-site safety efforts
- Add as-built and in-field model information

- Develop better recovery schedules through (5D) scenario building
- Use BIM for punch lists
- Prepare the model for project closeout

Model coordination planning efforts as outlined in Chapter 4, "BIM and Preconstruction," should define the tools, level of detail (LOD), and file formats that are acceptable to use during the construction phase of a project. The information exchange plan outlines how information will be transferred and audited. For example, is the architect going to give the model to the contractor to use freely during construction, or does an approval process need to be implemented? Additionally, the information exchange plan should state who will have ownership of the model and provide the necessary changes throughout construction to maintain the BIM files for the job. These issues, among others, are unique to a BIM project, where the transfer of data and the means and standards of exchanging data are of particular importance. As stated earlier, it is simply impossible for all the questions that arise in a project to be planned for and outlined in a model coordination plan. For this reason, the goal should be to provide governance as to how information will flow and accept that some issues will need to be resolved by the team as the project moves forward with a mind-set of flexibility.

Edit a Dimension? But Why?

At a national CAD user conference, at the end of a presentation on BIM, an audience member asked the speaker, "Drawing to real accuracy doesn't allow you to edit dimensions and dimension strings, though! If I have a 3 5/8″ stud wall and I dimension to the face of the stud, I want to round up to the nearest inch. Why would anyone want to model all of their dimensions exactly, if BIM doesn't allow for dimension editing?"

The speaker asked the attendee where he believed those dimensions in a string of walls went that he rounded up to. The attendee then responded that they were just included in the contractor's tolerances and was convinced that the contractor wouldn't construct something to the nearest 3/8″. The speaker responded by saying, "First, if I have a string of 10 walls and all of my 3/8″ dimensions are rounded up, that then means the last wall will be a total of 3 3/4″ short. Now if this is an ADA hallway, required clearance, or, worse, the boss's office, I don't want to have to explain to him why his office is smaller than the others because he was on the end!" The speaker went on to explain the importance of modeling exactly as you want the structure constructed and that avoiding coordination issues was much more important than "clean-looking, completely inaccurate dimensions."

Model Coordination

Model coordination is not limited to clash detection. In fact, the more models leveraged and used to understand and simulate actual construction, the better. The following examples are a sample of just some of the ways models can be used in the construction process. The term *model coordination* often refers to the use of models to coordinate or simulate some portion of construction. Whether this is systems coordination, scheduling, takeoff validation, model overlays, or other uses, coordination in this context means using BIM data to better inform physical work product to follow.

BIM and Site Coordination

BIM and site coordination can involve sustainable site management, building component tracking, commissioning, GIS, GPS material location, scheduling, and so on. Because this is a broad field and there are many tools for site coordination, in this section I'll analyze how BIM can be used to assist in organizing work in the field. Site work is unique because in many cases using BIM technologies goes further into more specific technologies that assist in a more coordinated project and integration (see Figure 5.1).

Figure 5.1 Example of a 3D site logistics plan

Site coordination refers to the organization of the site, materials, equipment, safety, and site security. Earlier in this book I outlined how to create a site logistics plan. Now I'll show you how to put it to use in the field. You can use this plan for many purposes, from safety controllers such as OSHA, to governing authorities, to subcontractors, to material providers. The site logistics plan will play a vital role in creating better communication and a safer project; the plan is often housed in the job trailer and posted on the wall or accessible on a project dashboard for reference. This assures that all team members have a visual understanding of where material lay-down areas, site accessibility, parking, and building access are. Although the site logistics plan is a static image of the site, there can be variations of the plan because of sequencing, scheduling, and different phases of the project. For this reason, it is not uncommon to have more than one logistics plan throughout the course of a project.

Using the site logistics plan is especially critical when the material being delivered on a job site is large or numerous. Combined with a tight site, such as in an urban setting, material lay-down coordination can be a full-time job depending on the size of the project. A BIM solution to this complex site coordination could be a sequencing animation or a series of site logistics plans. The question of what the material is and where the material is going can become a concern on-site as materials begin to pile up.

Current technologies have begun to leverage the use of RFID tags that can be placed on building components and scanned using a handheld computer, which brings up the building information about the component and where it is to be placed. This is particularly effective when dealing with complex structures or buildings that require multiple phases of construction. RFID tags can also be enabled with GPS locators. Paired with RFID-enabled software, the project manager can view where the building components are at all times. This technology, paired with a BIM model, provides material component information through the use of handheld computers and scanners as well as the location of components through GPS. This technology may also be used for construction equipment, cranes, bulldozers, and lifts as well. Virtually anything that a tag can be adhered to is able to be scheduled and located. As a result, using RFID and GPS tags on equipment is particularly interesting to construction managers who own and operate their own equipment. The ability to place an RFID tag on a piece of equipment can report information about the equipment including oil changes, routine maintenance, and issues to a field laptop instantly, which could potentially make the use of RFID tags more widespread in the construction industry. In the future, RFID tags could very well be tied to a web calendar that would send maintenance personnel reminders of equipment that needs to be serviced, and in turn, the maintenance personnel could identify where the piece of equipment is to do the work on it.

Site safety and security are important issues when dealing with any construction project. Although the use of BIM is important when there are potentially hazardous areas cited in the logistics plan, there is also technology, such as web-enabled security cameras, that go beyond the realm of BIM. Some construction projects provide the use of security cameras on-site to not only be provided as a safety measure but to analyze the progress of the structure for project team members to view over the web. Currently, there are a large number of security camera companies in the industry, yet the functionality of being able to move and position the camera remotely allows for functionality that didn't exist in earlier decades. This additional degree of visualization is a way for other stakeholders to gauge progress, potentially foresee any constructability and safety issues, and address them using this tool over the Internet. Some companies find it interesting to overlay both the BIM and the current construction snapshot views and compare the actual to the virtual level of construction completion as a way to see how and if the project is staying on track.

Clash Detection

Model coordination is where it all started. The ability for a construction manager to completely remove the light table from the equation and use 3D models from each scope and have the computer analyze the systems against each other based on prescribed rules was game changing. As discussed earlier, model coordination technology allowed contractors to be active participants in the design process for the first time.

In the next few exercises you'll see how to import files and test them using against one another. Additionally, you'll add a scheduling component to show how sequencing conflict reports can be created.

Navisworks Conflict Exercise

Autodesk Navisworks (www.autodesk.com/navisworks) is a powerful tool for construction managers using BIM. Navisworks is collaboration software that allows a design team to share, combine, review, and find solutions to correct a BIM model and 3D files using a 3D viewer. Navisworks can open multiple 3D files and combine them in a single workspace. Navisworks or similar software, such as Tekla BIMsight (www.teklabimsight.com), Bentley Navigator (www.bentley.com), and Solibri Model Checker (www.solibri.com), have similar functionality and can provide insight into the variety and abilities of software systems available in the construction industry.

Keep in mind the point of generating clash reports is to coordinate the model as close as you can to construction. What I mean by this is that the level of effort required to create and resolve model conflicts when a team is still designing the structure should be less than the effort required to coordinate very detailed model information coming from stakeholders, such as steel fabricators (see Figure 5.2). Additionally many subcontractors, such as fabrication and sheet metal shops, already use 3D modeling software that generates information you can integrate into a BIM workflow to streamline your processes.

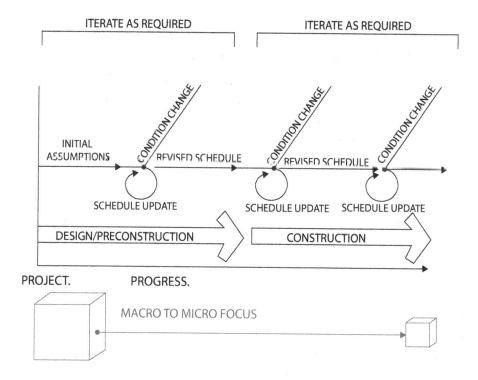

ITERATE AS REQUIRED   ITERATE AS REQUIRED

INITIAL
ASSUMPTIONS   CONDITION CHANGE   REVISED SCHEDULE   CONDITION CHANGE   REVISED SCHEDULE   CONDITION CHANGE

SCHEDULE UPDATE   SCHEDULE UPDATE   SCHEDULE UPDATE

DESIGN/PRECONSTRUCTION   CONSTRUCTION

PROJECT.   PROGRESS.

MACRO TO MICRO FOCUS

Figure 5.2 Conflict resolution path

Take the Macro to Micro Approach to Coordination

A common mistake contractors make when performing clash detection in the design review process is to try to resolve *all* of the issues they find in a clash detection report. Not only is this a waste of the team's time and efforts, it makes the process of resolution cumbersome and often doesn't achieve much because the design teams will shift and move components around.

Focus on delivering and resolving model conflicts in a "macro to micro" way. Start with resolving the larger systems, such as rooftop units, large ductwork runs, massive equipment, and other systems that would be improbable moving as design progresses because of their size and support structures.

Another mistake is inviting every stakeholder to all the coordination review meetings. Typically, clash resolution meetings only require representatives from the two systems being tested against and the model owner per your model coordination plan. It's a much better use of everyone's time and energy to involve the appropriate parties and to consider their time as a resource that must be managed. Be sure to explain to everyone

Continues

Take the Macro to Micro Approach to Coordination *(Continued)*

prior to having individual clash review meetings their purpose and function so as not to have anyone feel removed from the process. I usually say that anyone is welcome to attend another stakeholders' review meeting but the meeting will remain focused on the two systems being checked and analyzed. This leaves meetings open for others' involvement but focused on the review at hand.

Navisworks is not modeling software but rather analysis software. Although I don't cover it in this chapter, I encourage you to open Navisworks and in particular click the Open or Append function to review all the file types that the tool is capable of importing. The strength of Navisworks is its ability to import or link to a large number of model formats that are widely used in the design and construction industry today.

Additionally, I suggest that a user fully understand the differences between file extension types, such as .nwf, .nwc, and .nwd. The NWF file format tells Navisworks where to look for an updated or cached file location. These links can be in native file format such as DWG or RVT or Navisworks files, and they allow for easier updating in the clash detection review process, particularly when dealing with a large number of files to review. The NWD file type saves all loaded models, environments, views, and other input into a single version of a file. Also known as a "published" file format, NWD is a snapshot of a project. For this reason, this book uses NWD file formats in the exercises to eliminate file referencing. In the following exercise, you will use Navisworks to compile models into a composite model and test one model's geometry against another's.

What Is Composite Modeling?

Composite modeling is a modeling compilation strategy that combines the available 3D information into a single shared file. Composite modeling is not necessarily the ability to house all team members in the same office, developing the same model, using the same software. Although some companies are capable of this type of model development with architects, engineers, and contractors in house, it is unusual. Some owners who have a fast-track project using BIM find that one way to rapidly advance a project is to have a *BIM pit* or *BIM huddle*, in which all the members of the team, even those from different companies, have their office in one location where they work together to model and virtually construct the proposed structure. However, having the design team use a singular composite model is more common. A composite model is a series of 3D models that are created from the same or different pieces of software and that can be compiled for analysis and advanced visualization. Arguably, the most robust tool in which these models are compiled and tested is Navisworks.

Search Set Exercise

To start, you need to define the systems you want to test against. Models can be compared against each other in the following ways:

You can compare files against each other. For example, a structural model can be compared against a mechanical model. Although this method is feasible early on, it tends to generate a large number of "dummy or false" clashes because of duplicate model data. For instance, a floor slab in each model will create a conflict.

You can use search sets. Navisworks includes a search tool that allows the model components within Navisworks to be searched and grouped based on the name type. This is usually the most commonly used because it allows for granularity and for flexibility when models are updated.

You can use selection sets. Model components can be selected individually and saved as a set. These sets are useful to quickly determine if components in a certain area are in conflict with each other, but they are not the best means of comparing model information because they rely on the user to select each element correctly.

In this exercise you will use search sets to define the rules and parameters that you want to test. In particular, you want to test the ceiling heights against the ductwork to ensure you have adequate clearances. As similarly named items are added later, it finds these new elements with the same specified search parameters and links them automatically. This example uses the files Example-50% dd.nwd and 50% dd mech.nwd that you can download from this book's web page: www.sybex.com/go/bimandcm2e.

To open and append the model files, do the following:

1. Once you have downloaded the file, click the Navisworks icon in the upper-left corner of the screen and select Open. Navigate to the file's location (see Figure 5.3). Select the Example-50% dd.nwd file first and click Open.

2. Once the Example-50% dd.nwd is opened, you need to append the 50% dd mech.nwd mechanical file. To do so, click the Append icon located on the left side of the Home tab bar (see Figure 5.4). Browse to where you downloaded the 50% dd mech.nwd file and select Open.

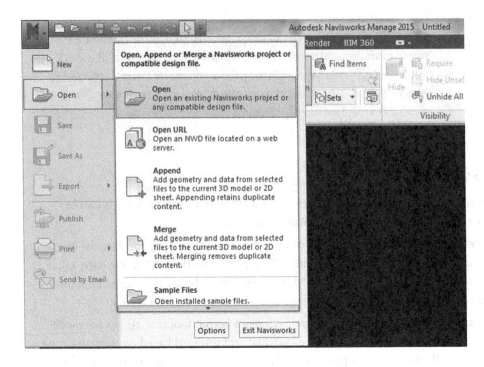

Figure 5.3 Open the sample NWD file.

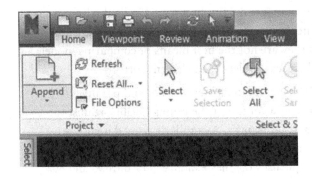

Figure 5.4 Appending the NWD mechanical file

3. Once both files are opened, select the Clash Detective icon in the Home tab bar (see Figure 5.5). This will bring up a new window in your Navisworks browser. You may wish to "pin" the Clash Detective window by clicking the pushpin icon in the top right of the window.

Figure 5.5 Opening the Clash Detective window

4. You can now begin to create a search set to compare against. In the Clash Detective window, you can see that there are many different selections you can make to compare against. Since you are concerned in this example with comparing the ceilings with the newly designed ductwork, you will create two new search sets: one for the ceiling and one for the ductwork. To begin, click the Find Items icon toward the middle of the ribbon on the Home tab (see Figure 5.6).

Figure 5.6 Opening the Find Items window

5. In the Find Items window that opens, you can build your first search set, which will be the ceiling. Click the plus icon to the left of the Example-50% dd.nwd filename and then click the plus icon next to Level 2 (see Figure 5.7). If you scroll down to the bottom of Level 2 you will see an item set named 2′ × 4′ ACT System. In this example, you want to isolate the ceilings in the building. To do so, go to the right and enter the values exactly as shown in the search set criteria (see Figure 5.8).

Field	Value
Category	Item
Property	Type
Condition	Contains
Value	2′ × 4′ ACT System

6. When you have finished entering the search criteria, highlight the Example-50% dd.nwd filename in the Find Items browser and then click the Find All icon at the bottom left of the Find Items window.

Note: Sometimes you can see the objects you've highlighted and sometimes you can't. It all depends on your view settings within the Navisworks browser. An easy way to check is to open your Selection Tree window and see if the items you searched for are highlighted in blue. If they are, you should be good to go!

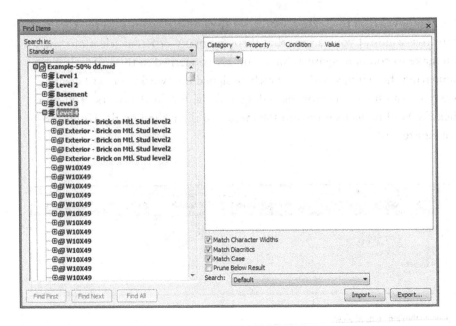

Figure 5.7 Opening the search selection

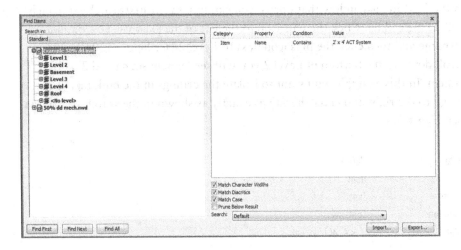

Figure 5.8 Entering the Find Item search criteria

7. Although you may or may not be able to see the model components highlighted depending on your settings, if you followed the steps correctly you have selected the desired model components. Next click the Save Selection icon and name the search set **2×4 Ceilings** (see Figure 5.9).

8. To isolate the ductwork in the model and create a search set, repeat steps 4 and 5, but this time select the category, property, and condition and then manually

enter the following value criteria in the Find Items window to isolate the
ductwork that you want:

Field	Value
Category	Item
Property	Type
Condition	Contains
Value	Design

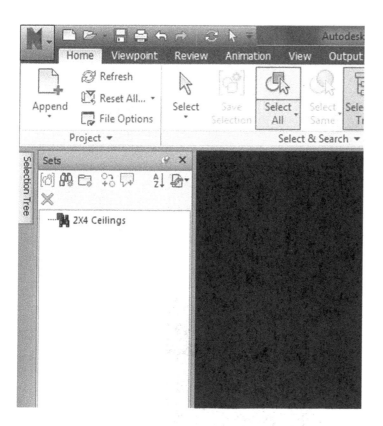

Figure 5.9 Saving your search set

9. Highlight the 50% dd mech.nwd filename and select the Find All function from
   the Find Items menu (see Figure 5.10).

10. Click the Save Selection icon and name this search set **50% DD Ductwork**
    (see Figure 5.11). Close the Find Items window, and save your file as
    Clash Tutorial.nwd.

Now let's move on to running a report between our two search sets.

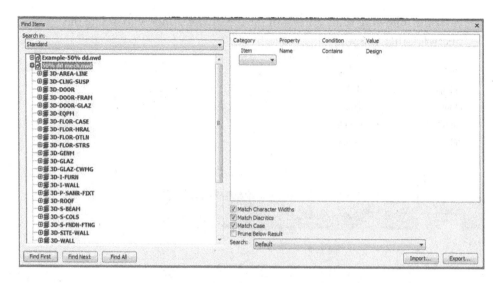

Figure 5.10 Creating the ductwork search set

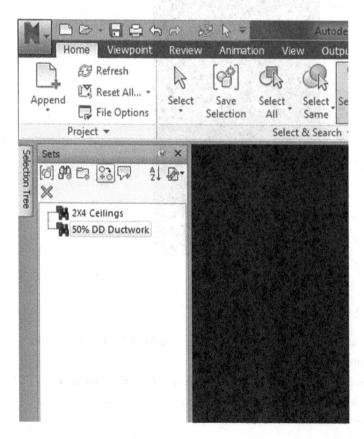

Figure 5.11 Saving the ductwork search set

You can create search sets for all the listed and available categories in Navisworks, which helps delineate one component from another more easily by letting the computer do the hard work of finding each component that meets your search criteria.

Clash Detection Exercise

You will now use the search sets you just created in your Clash Tutorial.nwd file to complete a clash detection report:

1. Open the Clash Detective window and dock it by clicking the pushpin at the upper right.
2. Navisworks defaults to a new test, but if needed click the Add Test icon.
3. On the Select tab, click the Selection A drop-down and choose Sets (see Figure 5.12). In the Selection A field, select 2×4 Ceilings.

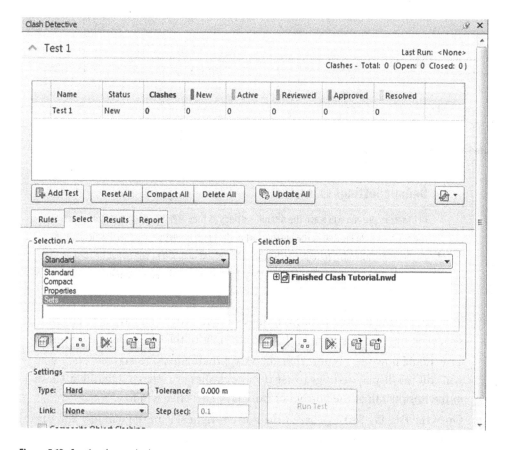

Figure 5.12 Creating the set criteria

4. Select Sets from the Selection B drop-down and choose 50% DD Ductwork (see Figure 5.13).

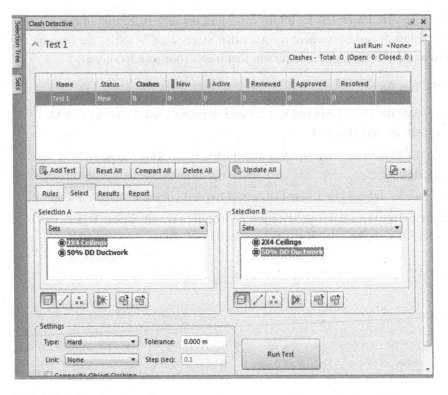

Figure 5.13 Defining the set comparisons

Default Settings vs. Custom

In this example you will leave the default settings as they are, but you may want to edit the Settings field located in the bottom left of the Clash Detective window. Particularly if elements are intended to touch each other, the default of 0.000 m will generate clashes for any items that are touching, even if they don't overlap and create a conflict.

5. You will use the default settings in this example. After you have defined your sets, click the Run Test icon at the bottom right of the Clash Detective window.

6. You should now see a report that shows approximately 716 clashes. Next you will edit the display settings to see if these clashes are related. Highlight Clash 1 in the Results tab at the bottom of the Clash Detective window.

7. Open the Display Settings bar at the bottom-right side of the window and select the check box Highlight All Clashes (see Figure 5.14).

8. You are now going to undock and move or minimize the Clash Detective window to look at the clashes. Using the Walk or Orbit view navigation tool (I prefer the Walk tool), move your viewpoint to the bottom of the ceiling. You can see here that you have an issue with the majority of your ductwork being able to fit in within the ceiling space (see Figure 5.15).

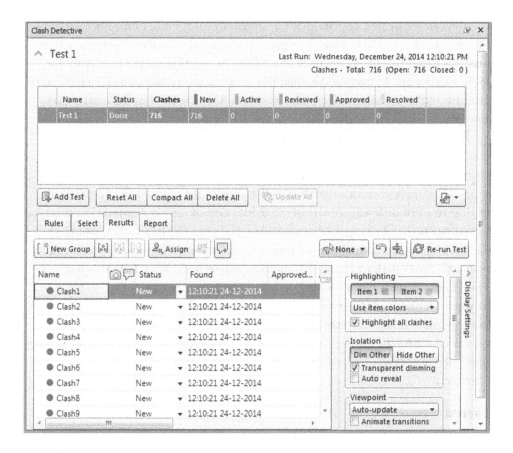

Figure 5.14 Editing the clash report display settings

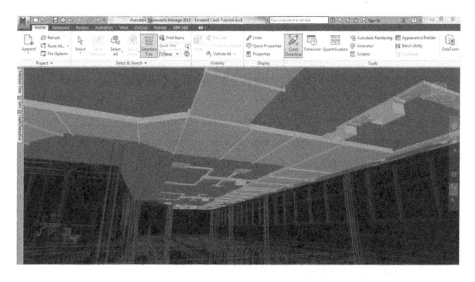

Figure 5.15 Identifying a system problem through a clash trend

You can see by this exercise that the larger issue is the size of the ductwork and the height of the ceiling. Instead of going through each clash in this scenario one by one, it is better to address the issue as a whole, or "one big clash," to resolve our issues in this design. As discussed earlier, this is one of the reasons why it's much more efficient in conflict resolution to look for trends and resolve them at the macro level first rather than looking at individual issues.

Future industry trends will decrease the amount of effort required in clash detection—specifically, the rise of cloud-based collaboration and model development platforms such as BIM 360 Glue and other custom-built virtual server environments that virtualize applications and let users collaborate in the same environments at the same time from all different corners of the world. One of the big disruptors in this type of technology is that it changes what's possible. For example, let's look at Autodesk Revit. Revit uses a check-in/check-out content process that ensures people working on the same file aren't making changes to the same model item. What's great about using this tool in the cloud is that now applications are available on the same cloud that tell users when they have modeled something that is creating a conflict, or "clash." This technology could reduce the number of clashes that a construction manager needs to resolve or eliminate the process altogether with integrated teams. This is an interesting example of where newer technologies supplant previous tools and processes in an effective way.

Fabrication

The use of BIM in fabrication has changed the way many parties in the construction industry operate, and it continues to have the potential to change the industry holistically. Since a BIM file contains parametric modeling information and many fabricators use 3D models to build their components, there is a great opportunity for information exchange between the two. These are some examples where 3D fabrication exists:

- **Structural steel:** Columns, beams, rebar, joists, and bracing
- **MEP components:** Ductwork, conduit, equipment, piping
- **Precast concrete:** Custom shapes, patterns, and reinforcing
- **CIP concrete:** Reinforcing layouts, component numbering
- **Specialty items:** Custom handrails, brackets, and sunshades
- **Glazing systems:** Customized assembly systems, connections, details
- **Other items:** Furniture, signage, site features, sculpture

As it pertains to a construction manager, the ability to receive shop drawings from BIM-enabled subcontractors during the documentation stage and compile and test that information in a construction model is an incredible resource. Although the field of subcontractors capable of producing fabrication models continues to grow, some haven't embraced this technology yet. In reality, subcontractors have the most to gain in a BIM process through the automation of production using computer numerically controlled (CNC) machines, which reduces up-front coordination time and equates to a reduction in rework overall. Additionally, the fields that offer BIM-capable applications grow in areas that require off-site fabrication. In the book *BIM Handbook: A Guide to Building Information Modeling for Owners, Managers, Designers, Engineers and Contractors* (Wiley, 2011), the authors explain the difference between made-to-stock (drywall, fixtures, studs), made-to-order (windows, doors, hardware), and engineered-to-order components (structural, MEP, custom concrete, and specialty items). Overall, BIM is most applicable to the engineered-to-order (ETO) components of the three categories.

The ETO category includes items such as structural, mechanical, and specialty items that are unique in design and construction to a particular project. Although they may be built using standardized parts and connections, the actual layout and design of these components are unique. Engineered-to-order items must be designed, sized, tested, analyzed, fabricated, and installed to exacting standards. The coordination of these components is often the most critical element of a project because the components deal with a number of relative unknowns through custom fabrication and because they can affect multiple systems. For example, a custom-angled curtain wall system will have many more unknowns than a standard rectilinear storefront system. Components such as customized supporting structure, custom flashing and waterproofing, and custom-fabricated mullions all are systems that require unique solutions to be resolved.

The proficient utilization of BIM coordination during fabrication carries into construction. With this in mind, it is best to analyze the BIM files prior to construction to reduce the issues in the field. Much of this can be achieved by completing a clash detection analysis with other systems or a sequencing clash (to verify that the size of the preassembled components can be installed) by using the fabricators' file. Because the fabricator's model is a true representation of what is going to be sent to the machines to construct, it is the most accurate model a team can use. This BIM-as-built approach to documentation is a logical progression of the virtual design and construction (VDC) BIM. As you increase the information and integrate more detailed models, you increase the accuracy of the project and reduce overall unknowns.

The concept of BIM-as-built directly applies to lean construction coordination prior to execution. Lean production and procurement are parallel efforts because BIM fabrication is enabling the following to take place:

- Offers the ability for the construction team to procure and fabricate components accurately (earlier if need be)
- Limits the amount of rework and hours in-field
- Reduces paper waste because of digital submittals and coordination

- Improves team buy-in and trade coordination
- Reduces overage on ordering and waste
- Reduces total costs for engineering and shop drawing time

Obviously, using BIM for fabrication lends itself to an integrated process because it simply wouldn't be possible without the early involvement of the subcontracting/fabrication team. Although different opinions about the direction of model coordination exist, the overall thinking seems to be that the process of refining the VDC BIM to a fabrication level of detail is most effective (Figure 5.16) prior to commencing construction.

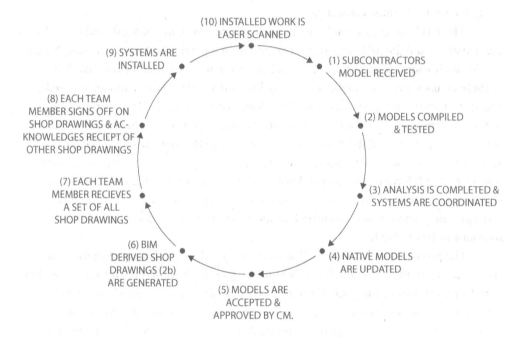

INSTALLATION VERIFICATION WORKFLOW.

Figure 5.16 Refinement of BIM from design to fabrication

Although some of these processes might overlap, such as constructability and engineering reviews, the overall refinement of the VDC BIM to a fabrication model is developed in tandem with the actual project design. For example, the tools that an engineer might use to test a shop model such as Tekla or SDS/2 for structural and IES or eQUEST for mechanical are tools specific to that team member. And it's important that an engineer be focused on verifying the performance of the systems and use the tools available to complete this work.

Other analysis such as constructability issues where BIM can be leveraged for clash detection, clash sequencing, increased schedule visualization, clearance clash detection, reverse clash detection (deviation testing), and other trade coordination can occur in tandem with the engineering analysis, though on separate software. Both types of testing and analysis are completed and overlap for time's sake; however, each have different results sent to the project team. This evolution of the model's level of detail and the refinement from analysis lead up to the construction phase of a project. In the field, the resulting product should be a usable, accurate tool, which can then be used to verify the accuracy and installation of the entire project.

The level of detail a subcontractor and fabrication team are capable of allows the construction manager to associate even more advanced 4D scheduling capabilities to the models to further refine and coordinate the construction. Because the building has been constructed virtually, the ability to produce more detailed scheduling coordination is possible as well.

Navisworks Sequencing Conflict Exercise

Navisworks can also create clash detection reports tied to a "simulation," or schedule. These reports are extremely valuable when different construction is going to take place concurrently, such as a medical room receiving a large piece of equipment such as an MRI machine in which the room needs to be built around the piece of equipment. The ability to visually validate and run a sequencing test to determine if schedule assumptions require refinement is extremely valuable. Another example may be showing when safety or scaffolding equipment needs to be removed for construction activities to occur on time. Although construction scheduling animations tell the story as an overview, the sequencing clash function allows the construction of such a facility to go more smoothly.

In this exercise, you will create a sequenced clash detection report to test the timing of the concrete pour for a housekeeping pad prior to that of the main slab that requires curing in the basement of your facility:

1. Open the construction-sequence.nwd file in Navisworks Manage from this book's web page.

> **Note:** Notice in this file there are two models: arch-model and housekeeping pad. This exercise simulates two concrete pours in the basement of the proposed building. So, in this example, you have a slab on a grade that needs to be poured, and you need to verify that the curing time—which in this example is 30 days—doesn't overlap with the timing of your new pour for a separate and isolated housekeeping pad.

2. Open the Clash Detective window.

3. In the left window, select the 6″ Concrete Basement Floor item that appears under arch-model.nwd ⇨ Basement.

4. Select housekeeping pad.nwc on the right (see Figure 5.17).

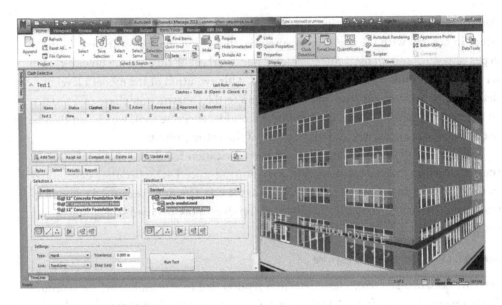

Figure 5.17 Establishing the model clash parameters

5. Change the clash type to Hard, and change the tolerance to 0ft 0.00.

6. Change the Link setting from None to TimeLiner, and leave the Step (Sec) setting at its default (.1).

Note: If you'd like, you can investigate the embedded schedule by clicking on the Timeliner and reviewing the basic schedule this example uses.

Note: In this example, you're using the Hard clash type because you are making sure that any model components abutting or embedded in the slab floor do not intersect. This type of clash may also be done using the Tolerance tool, with a user-defined tolerance setting to verify clearance between two objects as they are constructed.

7. Click Run Test.

8. On the Results tab, view the clash and highlight it. (This example should have found one clash.)

You will see that you have a conflict in pour times between the slab floor and the housekeeping slab. Additionally, you can see that the date of the clash appears in the upper-left corner of the model view window and that the Timeliner window's Simulate tab has already advanced to where the clash is occurring (see Figure 5.18).

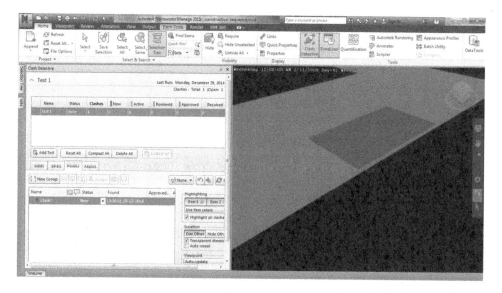

Figure 5.18 A clash is generated because of a schedule conflict.

Although this is a simple example of how sequencing clash detection works, it shows the value in being able to verify for complex structures that critical path construction elements aren't interfering with one another. As construction progresses, it is also a means of viewing which components are to be installed and in what order. Lastly, the sequencing clash can be used with mass model elements that indicate material lay-down areas, which can then be tied to procurement schedules to verify successful site coordination. The sequencing tool is powerful and, when used in tandem with a solid BIM strategy, can be used to better coordinate and create efficiencies in the amount of time it takes to review 4D construction issues. Interestingly, the construction community has used this technology as a springboard to more broadly look at areas where BIM could be used.

BIM Scheduling

Before the first shovel of dirt moves, the construction manager must do a significant amount of work. One of the most important tasks is preparing the construction schedule. The schedule is one of the driving factors in the success of any project and is a critical component to all team members as it provides guidance to where and when they should perform work.

At the onset of a project (and often before that), a member of the construction management team creates the schedule. This schedule typically reflects experience with construction timing: material lead times, weather, crew, and equipment concerns. The importance of a schedule in regard to BIM is to better inform the team and track progress from the beginning to the end of a project (see Figure 5.19). So, how does BIM improve schedule management? How can you use BIM to increase visualization and schedule accuracy?

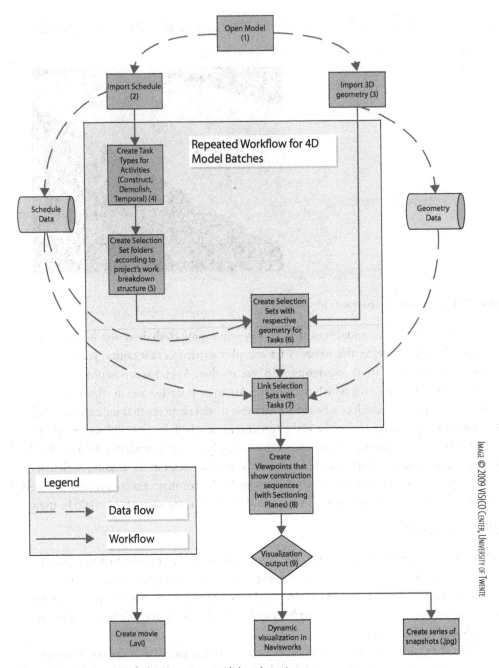

Figure 5.19 Updating a BIM schedule is a continuous task through a project.

A construction schedule is a sophisticated chart or table showing tasks and the times required to accomplish them (Figure 5.20). Although there is an implied correspondence between a task and a building component, there is no direct link between the drawings, specifications, and the construction schedule. As the design progresses, the construction manager reviews the updated drawings to identify changes in scope, as well as the addition of design elements, and then updates the schedule to reflect these design changes.

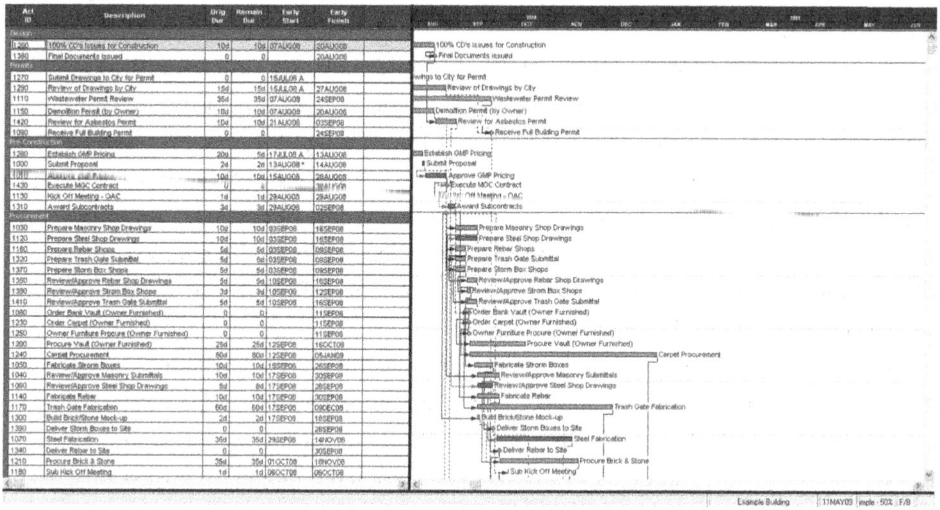

Figure 5.20 A schedule is a series of complex, overlapping tasks to ensure successful project delivery.

The refinement of the schedule relies on the accuracy of the construction manager to review the new design documents each time and judge the projected availability for additional equipment, material quantities, and so on. The schedule, and the subsequent revisions, is one of the more time-intensive aspects of a project, and the members of the team rely on its accuracy to deliver a project to the owner on time. Therefore, any increase in efficiency and schedule accuracy can do two things:

Note: I mean *efficiency* here not only in terms of time but also in terms of costs, accuracy, and thoroughness.

- It can provide the construction manager with more time to further coordinate other tasks.
- It can mitigate many of the issues associated with schedule misinterpretation and accuracy through enhanced visualization by linking the schedule to the virtual construction.

One of the key goals of BIM is to virtually build the structure before physically constructing it, so the introduction of scheduling is a main component in BIM's effectiveness.

In an exercise later in this chapter, you will simulate real processes by using an example design model to generate a BIM *sequencing simulation*, or "4D" animation that is the link between the model and the construction manager's schedule. The use of sequence simulations is an extremely valuable tool—you'll learn more about this throughout this chapter. When selecting a BIM schedule simulation tool, it is important to consider the ease with which model components may be added and as the schedule changes, the work associated with updates while still providing a robust resource.

A *sequencing simulation* shows in 3D the building being built from start to finish; it helps communicate completion dates to owners, gives subcontractors and tradespeople a better understanding of the scope and timing of their work, and helps field personnel verify that the project is on track. To use BIM as a tool for scheduling, the model-sharing language and accountability discussed in Chapter 2, "Project Planning," must be in place. If language hasn't been established, there may be some challenges for the contractor to receive and use a designer's BIM file, and there may be issues associated with the LOD in the model.

Although there is no need to delve into great detail about sharing the model during this phase, contractors should state their intentions for using the model in the BIM execution plan. Additionally, the construction manager must understand that a design development-level model is by no means a completed model. In a BIM process, it is helpful to establish an understanding that model sharing is critical to accomplishing more integration, especially if the construction manager is to advise the design team through the preconstruction phases.

After the model exchange parameters have been established, you can begin creating a scheduling animation.

The advantage to working with your design team's model(s) is threefold:

- Early coordination exchange is facilitated in the design phase. Users who are experienced working with building information modeling can begin asking high-level questions early to make sure sequencing considerations are being made.

- There is a cost savings in not allotting additional resources to remodeling a structure each time. Creating a secondary construction BIM model is typically a waste of resources in a project and introduces risk in versioning control.

- The team works from one single source of truth. As discussed in Chapter 2, the BIM execution plan delineates model ownership throughout the process. By working from a single point of model issuance, the process of coordination and control is simplified and revisions can be made much more clearly.

Although these are all benefits, the main benefit to beginning with the architect's model is that you're using the product developed by the design team, which is a best practice. Many times, the response to an early model provided by a designer to a contractor is that the model is not complete enough and that "I am going to have to create my own model instead." A model typically changes many times prior to creating the construction documentation and continues to grow in detail as the project progresses. Starting a separate model will not introduce any efficiency into a BIM process, and it is better to use a single model and inform the architect of any big issues associated with their model as opposed to putting it by the wayside. Using the designers' model helps to identify new items and scope as well as coordinate owner-driven design and program shifts—in other words, coordinating once instead of twice.

Scheduling Software

Construction scheduling software, such as Primavera (www.primavera.com), Synchro (https://synchroltd.com/), or Vico (www.vicosoftware.com), keeps track of the work breakdown structure (WBS) and critical path dependencies between overlapping tasks to create complex timelines, which can be displayed in a variety of standard formats. The software can be used for planning as well as for tracking projects once they are under way. Schedules are constantly updated, and the software helps update the project's schedule.

Both Microsoft Project and Primavera systems are compatible with Navisworks TimeLiner as well as customized Microsoft Excel (.csv) files. Navisworks will also read other scheduling software that can produce an MPX or a Primavera file. The following exercise uses Primavera to demonstrate how to link a static schedule to a BIM schedule. The power of most scheduling software is its ability to easily overlap, link, and create very complex schedules with large amounts of tasks tied to a timeline. Updating these schedules can be a constant source of work that is required to define the progress of a project. These scheduling programs and others simplify the task of creating these complex schedules and are commonplace in the industry.

Note: Remember when working from Primavera that future revisions and changes supersede the old schedule. Always archive your old schedule, and save over the old schedule with the same filename.

This book shows how to use Navisworks to run a schedule simulation, sequencing clash analysis, and clash detection. Navis has other tools, but these three are the ones used most by construction managers. The greatest benefit to using Navisworks is the ability to combine several files of many different file types. Again, Navisworks Manage is not a modeling program; rather, it is an analysis program that links BIM and 3D files into a Navisworks format (NWD). Navisworks Simulate and Freedom viewer are useful for those who might want to look at conflicts or at the composite model overall but who don't want to purchase the full version or any licenses of Navisworks.

As mentioned, the purpose of a scheduling animation is to show in three dimensions over time the building being built from start to finish. Ultimately, the quality of the animation is directly related to the quantity and accuracy of the model components. Keep in mind that it requires additional time to link more components to the schedule. Whether that involves the creation of additional search sets for model components, or whether the granularity in the schedule is quite detailed, the more detailed the information, the longer it takes to link to more lower-level schedule elements. In a sequence simulation, you can show the earthwork excavation, site demolition, pile driving, piers, excavation, forming, site utilities, crane erection, truck loading areas, staging and lay-down areas, reinforcement and rebar, concrete foundation pour, structural steel erection, and so on. Almost any activity that occurs during construction can be modeled and simulated if represented by a virtual model component. With Navisworks, you can create detailed or simple animations using these 3D model components.

To begin this example, you will learn how to import a Revit model file, just as if you were receiving it from a designer:

1. Download Example-50% DD.rvt from this book's web page.

2. Launch Navisworks Manage, click the top-left Navisworks icon, and click Open (see Figure 5.21).

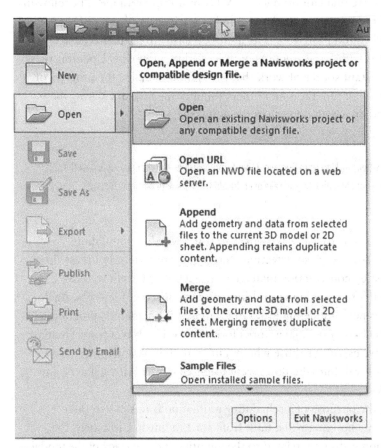

Figure 5.21 Opening a Revit file into Navisworks

3. Navigate to where you downloaded the example file and click Open. The file may take some time to load.

Default Units in Navisworks

The default is metric in Navisworks, so I usually find it helpful to change the Unit setting to the corresponding file units. To change your units, click the Navisworks icon in the upper-left side of the screen and click the Options button at the bottom right of the window, as shown here:

4. Maximize the interface tree and select the Display Units line (Figure 5.22). Here you can edit the units to your preferred settings.

Note: It is important to note here that you are saving the file as an NWF file in Navisworks. This will keep your link to the model that you opened active and will refresh the model(s) loaded into Navisworks each time you open or refresh the file.

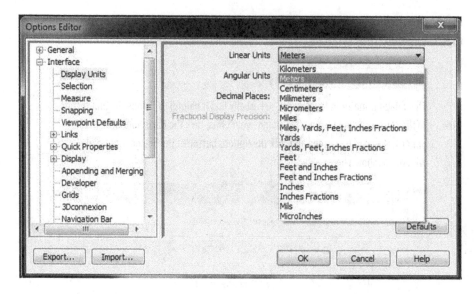

Figure 5.22 Changing the file units in Navisworks

5. Click the Navisworks icon again and choose Save As. Select the NWF file type option and click Save.

Now you'll change the Snap settings in Navisworks:

1. Navigate back to the same Options window as earlier for the Units selection and choose the Snapping option.

2. Select Snap To Vertex, Snap To Edge, and Snap To Line Vertex (see Figure 5.23).

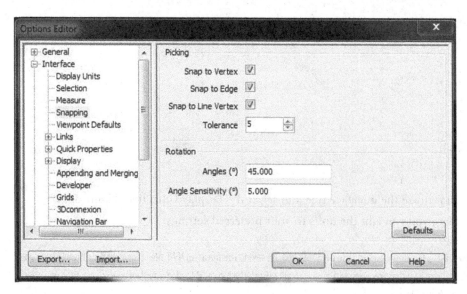

Figure 5.23 Enabling snaps in Navisworks

These changes will make selecting objects easier and can be edited to the user's preference.

Creating a 4D Simulation in Navisworks

Now that the model is saved in Navisworks, you need to import a schedule for linking. In this example you will use the file construction-model.nwd and the existing Microsoft Project schedule Commercial Construction.mpp (available on this book's web page, www.sybex.com/go/bimandcm2e). Download the file Commercial Construction.mpp. You will link this simple schedule during the conceptual stages of a project and add detail later.

To import the schedule into Navisworks:

1. Click the TimeLiner button on the toolbar. The TimeLiner window opens at the bottom of the screen.

2. Click the Data Sources tab, and click the Add button to open a context menu.

3. Choose Microsoft Project MPX 2007-2013 (Figure 5.24) and link this to the MPX file named Commercial Construction.mpp and click Open. This opens the Field Selector dialog box.

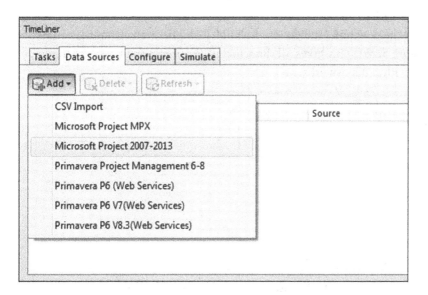

Figure 5.24 Linking the MPX file

4. Once the schedule file has been input into TimeLiner, click OK in the Field Selector (Figure 5.25) to accept the default. This should add the link to the TimeLiner window.

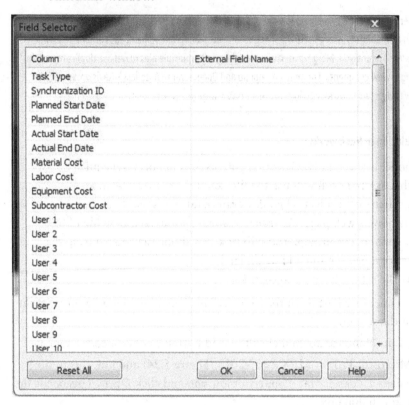

Figure 5.25 Accepting the default import settings

5. Right-click the new link, and choose Rebuild Task Hierarchy from the context menu (Figure 5.26). This takes all the schedule line items and breaks them out into tasks within Navisworks.

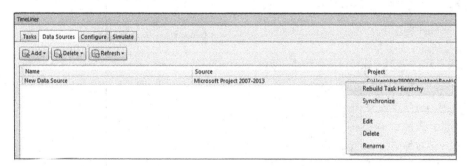

Figure 5.26 Rebuilding the task hierarchy from the link

6. Click the Tasks tab. All the line items in the schedule are now tasks, with start and end dates.

You may now begin linking tasks to model components. You can go about this in a couple of ways. As mentioned earlier, a model search set is typically the easier way to link schedule items in the model. As similarly named items are added later, it finds these new elements with the same specified search parameters and links them automatically. Let's create a search set named "Footing" using the following search parameters from the clash detection exercise earlier in this chapter (Figure 5.27).

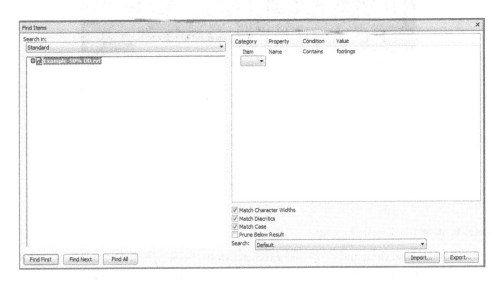

Figure 5.27 Creating the Footing search set

Field	Value
Category	Item
Property	Type
Condition	Contains
Value	Footing

Again, search sets look within all the listed and available selected categories in Navisworks, which help delineate one component from another more easily.

Once you've created the search set in Navisworks:

1. Name the search set **Footing**.

2. Once you've created the search set, scroll down the task list in the TimeLiner window to the Task activity labeled "Pour concrete piers and foundations" under the Foundations header.

3. Select the Footing search set from the Sets And Manage Sets window (below the Find Items icon), right-click, and select Attach Current Search from the context menu (Figure 5.28).

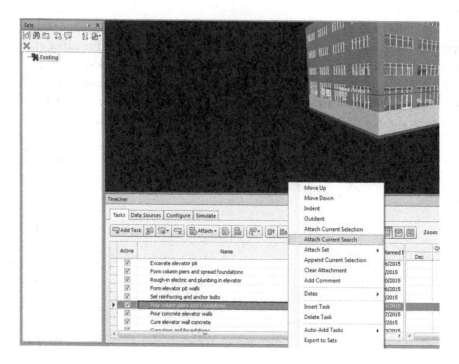

Figure 5.28 Attaching search sets to the schedule

You may also drag and drop the search set from the Selection Tree window to the task. The updated status is reflected on the "Pour concrete piers and foundations" line in the listing, indicating the search has been linked to the task successfully.

4. Scroll to the right of the TimeLiner window and click the Task Type field in the same line. Select Construct from the drop-down list to indicate that these are construction activities rather than demolition or temporary (such as shoring or formwork) (Figure 5.29).

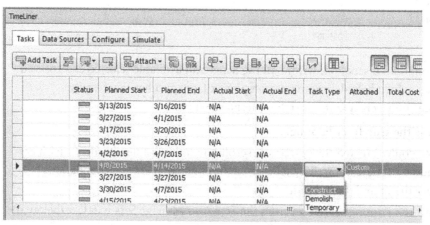

Figure 5.29 Defining the task type

5. Click the Simulate tab. Since this is Planned work versus Actual, click the Settings button, and select the Planned radio button at the bottom of the Settings window. Set Interval Size to 1 Percent and change Playback Duration to 60 seconds (Figure 5.30).

Figure 5.30 Editing animation settings

6. Still on the Simulate tab, click the Play button or manually drag the timeline slider to see the results (Figure 5.31).

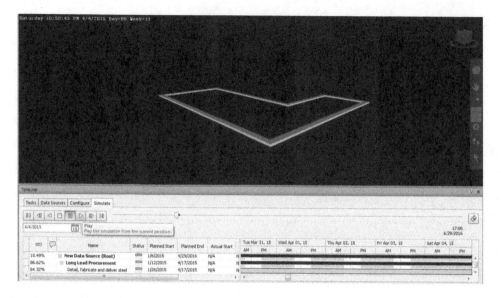

Figure 5.31 Running the simulation

7. Link the rest of the model to the schedule as desired. For further simulation editing, click the Settings button and configure your sequence as you'd like.

Using this powerful tool in Navisworks, you can simulate a schedule in 3D to better communicate the order and construction of a structure. Although this is an introductory exercise, there are many other features to explore and even more streamlined ways of creating 4D simulations. Additionally, Navisworks is not the only 4D tool on the market. Other tools such as Vico, Synchro, and ConstructSim can also achieve sequencing simulations. These tools vary in cost, however. Some of the more robust tools such as Vico integrate estimates and location-based systems (LBSs) into the 4D simulation. Solutions such as Synchro can produce rendered videos that give the simulation a more polished feel.

Regardless of the tool choice, the creation of sequence simulations takes a bit of effort the first time they are built, but updates usually take much less time to create. As you put together your 4D plan of attack, think of what it takes to update your simulation and streamline that work as much as possible. This will make updating throughout construction easier and more useful. Specifically, you can use a scheduling animation in the field to visually represent the degree of completeness for the project, which helps assign some visual basis of completion to contractors and subcontractors. In essence, BIM bridges the gap between model component and schedule and is an invaluable tool to the construction manager.

Completing the Feedback Loop

Construction has always had connectivity issues between the field and the office. In one environment, you have construction equipment, material deliveries, and subcontractors all moving around and supporting the work being put in place. On a daily basis, the environment and conditions change in the field as work progresses. Conversely, office staff is quite often off-site and usually supports an array of projects for a construction management firm at one time. The administrative support team assists a project on many fronts such as making sure paperwork is in place, legal, invoices, insurance, human resources, bonding, and so forth. So how does the office stay connected to the field? How can solid communication between many jobsites be a reality?

For starters, you need to realize that the disconnected nature makes communication difficult simply because of geographic disparity between each role. On projects where on-site personnel are able to support a project, many efficiencies can be gained, but for most projects this isn't a reality. So how do you use BIM and technology to begin to better connect the field to the office? It starts with building a *feedback loop*.

Feedback loops aren't necessarily a new thing, although they have typically not been a focus for many construction management firms as most of the attention is directed to where work is occurring; never fully realizing the efficiencies that can be gained by better connecting these resources. Here are some examples of where feedback loops are needed:

- Determining the amount (or percentage) of work put in place against a total subcontract value to determine the proper amount to bill. The field staff can visually verify the work, but the office needs to send payment for the correct amount.

- How many pieces of material are needed to make sure construction flows smoothly? For example, when is the steel fabricator able to deliver the first level columns? Quite often procurement is handled in an office, but the tools need to be delivered to the field on time to stay on schedule.

- How can you tell if you are on schedule? Often in construction you see work being done, but how can you tell if it is enough work and in the right areas so project managers can make accurate status updates to leadership?

- As issues pop up, how are you able to get them answered and resolved as quickly as possible? Is there a way to share better information, faster? Think of the time it takes to resolve an issue such as a large piece of equipment not being able to fit where it should or a subcontractor who has installed their system in the wrong location.

These are just a few cases where connectivity between the project and the office is critical. You have learned that BIM has a critical role to play *during* construction, not just in preconstruction, in addressing some of these feedback loop issues. Many professionals view BIM as a tool for only preconstruction, when in reality there is significant value to be gained by leveraging it effectively in the field. There are five major areas where BIM has a role to play as work progresses to complete the feedback loop: systems installation, installation management, installation verification, construction activity tracking, and field issue management. Let's walk through scenarios for each of these areas and discover how BIM can be applied.

Systems Installation

If you think of a model as a "kit of parts" that needs to be assembled in the field, there are major efficiencies that can be gained in the field through better visibility. I've shown how you can use composite models for conflict resolution between components, and you can also use these models as a more effective means of measuring what work has been put in place. For example, many construction managers use a different coloring system to reflect where components are in the supply chain, and once the composite model is constructed to a high level of detail, they have a very usable tool, which will be demonstrated in Chapter 6, "BIM and Construction Administration."

Think about it: If you take the approach that "if it is to be constructed, it must be modeled," the work can be better tracked through the model to verify installations more effectively. Certain tools, such as BIM 360 Field from Autodesk and Field Supervisor from Bentley, can be customized to automate the color-coding process. Although color coding may not necessarily be applicable to certain items such as site grading or landscaping (depending on your model), it should be able to handle the components modeled within a structure. At a high level, components may be broken down and color-coded as follows (example) to track project progress:

- Awaiting shop drawing approval—cyan
- Shop drawings approved, in fabrication—dark blue
- Fabrication complete, materials en route—orange
- Materials on-site, uninstalled—pink
- Materials on-site, installed—green
- Materials installed and commissioned—yellow
- Scope complete—gray

Combined with the use of a cloud-based collaboration tool or VDI/VDE environment that is mutually shared between the field and the office, both teams are able to explore, isolate, and analyze construction progress in the field to ensure accurate billing and forecasting more effectively. The use of construction field tracking software or custom application can be a lifesaver in these areas; as field personnel walk the project, they can update the model graphics faster. This technology should improve as a way of visually validating construction progress and as the industry continues to focus on more connected apps to their enterprise systems.

Installation Management

Material quantities, procurement, and installation in construction have never been an exact science. Typically as the construction process progresses, the information available for the number and amount of components becomes more detailed. Using BIM for materials management is a key component of completing a critical feedback loop that construction managers have to take care of.

In a typical project, a construction manager (who is not self-performing) will work with the subcontractors on a project to verify material availability and delivery on-site to perform work per the project's schedule. In many cases, materials can't be delivered all at once. A jobsite may not be large enough and unable to store all the materials for a particular scope of work. The production of a project's components is staggered and requires multiple deliveries to a jobsite, or some materials require climate-controlled storage conditions. These constraints are quite common, particularly on complex projects in urban environments. Additionally, for projects that are fast-track or using lean just-in-time delivery methods for materials to streamline construction, it is critical to understand how much of each material is needed to ensure good project flow.

Delivering Work Using LBS and Lean

Rex Moore is an integrated electrical subcontractor that is an early adopter of LBS and lean production. The work they do is now planned down to the day. In-field installers receive daily "kits" with the components, drawings, tools, and information workers need each morning and they are removed each night. This method of delivering work has made them more efficient and significantly more productive. Here's a description from the Rex Moore team:

Rex Moore's Production System Evolution

In mid-2009 we were starting our lean implementation and during brainstorming sessions the words "lack of standards" came up as a root cause for almost every problem we had. With that theme every major function of our value stream went back to their groups and began to identify standards that could be implemented. We understood, or at least trusted our organization's direction that without standards there can be no improvement. Design, manufacturing, supply chain, human resources, etc. began to develop what their standard process would look like based on the foundation that it must reduce waste and provide value to our internal/external customer.

One group was focused on design for manufacturing and install. In the past we usually went through a project's drawings and reviewed for constructability, material requirements, prefabrication opportunities, and estimated installation labor hours and produced detailed drawings. Not all projects followed this process and the ones that did had varying levels of deliverables. Furthermore our processes for installing electrical systems on-site varied from person to person.

We decided to do a Kaizen event and run a pilot Design for Manufacture and Installation (DFMI) project for Copart, a small datacenter in Reno. Rex Moore was the prime contractor, so we had more flexibility with the schedule to allow prototyping. The team reviewed another company's concept of material kits and unique identifiers for those material kits to get a vision of what could be accomplished. With that vision, the group began to develop standard installation and kitting methods for the pilot project. The team used brainstorming to get ideas written down and used affinity diagraming to group those ideas. Practical problem solving was used as a roadmap to keep the group on point, and process mapping was used during process development.

One of the major outcomes of the Kaizen event was the development of a process to sectionalize the project into a work breakdown structure (WBS). WBS is a consolidation of all work activities that must take place in order to consider the project complete. The lowest level of the structure and where the work is performed is a four-digit code that tells you where and what work is taking place on the project site. The structure of the code is standardized across the company and is never duplicated on the same project. The code was coined "Kit Number" and gave us a standardized way to communicate about project deliverables.

Project teams could now apply due dates to each kit number on their project, which would drive our design department to completing the detailing of a kit. A completed kit design would flow downstream to supply chain for procurement, then to manufacturing for production/delivery of the kit(s), and finally to the field, just in time for installation.

Having a standardized kitting process in place allowed everyone in the value stream to provide feedback on portions of the process that could be improved on or standards not being followed. Only after the process was standardized could improvements be made and deviations identified. Some of the major improvements that came directly and indirectly from the Copart Kaizen were:

- Standardizing on kit types, which allowed for a reduction in the installation methods and material contents of a kit. This reduction of methods and parts increased the installer's familiarity with kits, reducing lead time.

- Creation of standard manufacturing processes and design of standard assemblies that could be made to stock to reduce the lead time and quality of the manufacturing process

- Implementation of an inventory management system to reduce project-specific inventory levels into a shared inventory of like standardized items. Required dates and lead time drive the procurement dates, which reduce cash-out and use of warehouse space.

- Creation of standardized packaging and use of rolling cages to allow for easy receipt and mobility of kits on-site to the point of use, resulting in reduction of motion on-site

- Gaining visibility of true material requirements across all projects to allow for negotiation with our suppliers for the best price and lead time

Continues

Delivering Work Using LBS and Lean *(Continued)*

Although the goal of the Copart Kaizen was to reduce installation hours and increase profitability, the major benefits were not realized at a project level. The Kaizen team never planned on tracking cleanliness of the project, or safety incidents that never happened. However, a clean and safe project site was an outcome. The sum of many projects like this led Rex Moore to the lowest safety performance index ever, going from a consistent level of around 93 down to 63 after the implementation of lean concepts and tools. Standardizing the processes also allowed for an increase in revenue per employee as well as opening up opportunities in new markets with new customers that traditional methods would not have allowed.

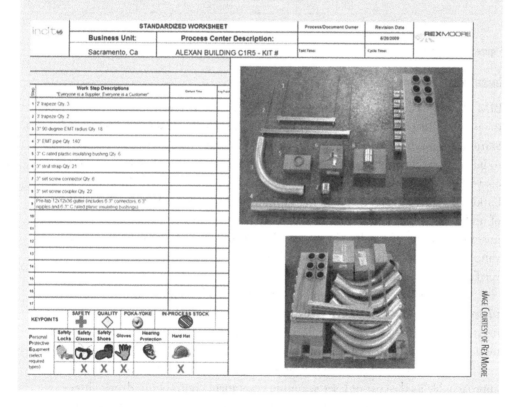

IMAGE COURTESY OF REX MOORE

BIM is a great tool for analyzing and determining the way you are going to approach a project. One of the best tools to develop a nondestructive model LBS plan is Vico Office (Figure 5.32). Vico uses bounding boxes to organize and simulate work using the composite BIM to develop the schedule. The outcome is a detailed materials breakdown and a production ordering schedule that optimizes just-in-time delivery.

This method of slicing and dicing the model to develop installation scenarios also allows users to optimize the construction schedule based on production rates, crew sizes, and quantities.

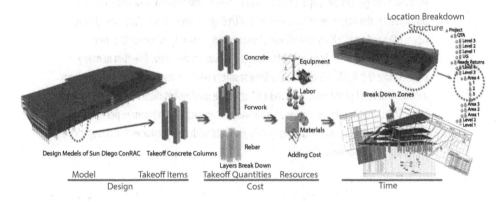

Figure 5.32 Vico Office logic

Installation Verification

BIM already has a role in installing work in the field. Currently, many systems support the use of models to lay out a project in the field. Using tools such as Trimble's Total Station, the model is imported and construction points are sent to users in the field to build from. Such tools are quite accurate and if used correctly allow for better control of digital information in the field, but they cannot remove the ability for a builder on-site to install a system inaccurately if they choose. Whether it's due to lack of understanding of routing around other systems, a "first in" mentality, or a genuine misinterpretation of the drawings or models, installed work needs to be checked and validated against. This installation verification is the final step in a typical installation verification workflow.

Installation verification can be accomplished in a number of ways. Some prefer to do it by in-field measuring devices and notepads, and some use cameras. But the most effective means of capturing in-field conditions and checking them against the model is through the use of laser scanning and overlays.

The ability for projects to use laser scanning is becoming increasingly approachable as the cost of laser scanning equipment continues to drop. Three-dimensional laser scanning is a process that uses a scanning terminal to capture high-density 3D geospatial data from the environment around it. Think of it like a panoramic picture in 3D that captures millions of points that represent the distance from the scanner to the object. Each scan collects millions of points that contain dimensional data for everything within view of the scanner. Each scan produces a "point cloud" that can be compiled with other point clouds, similar to a geographical survey (Figure 5.33).

Figure 5.33 Laser scanning team in-field

Once the final point cloud is created, it can be converted to a model or imported directly into composite model viewing tools such as Autodesk Navisworks, Bentley Pointools, or Tekla BIMsight for analysis. The value in the overlay is that teams can quickly determine whether or not the installed work is in tolerance (Figure 5.34). If it isn't, then the team can also determine what other systems will be affected by the inaccurate installation and begin to either coordinate or have the system reinstalled to avoid issues.

Figure 5.34 Laser scan and BIM overlay

Emphasizing Accuracy

As a best practice, it is usually a good idea to have each installing trade understand that they are to install their system as accurately as possible. I typically tell everyone that as a result of their participation in the conflict resolution process (mentioned earlier in this chapter) their systems are to be installed per their submitted model and shop drawings and will be checked over time against laser scans that will be completed on the project. Those who wish to deviate from those submitted models are at risk of installation conflicts and will either have to resolve them on their own nickel or reinstall them entirely.

This approach increases awareness that installations will be checked for accuracy fairly and frequently. Additionally, the use of the laser scanner over time or phased-based scanning is a useful means not only of comparing physical and virtual conditions for analysis but also of accurately collecting 3D as-built data for use in delivering the as-built conditions of the project downstream.

Construction Activity Tracking

Construction activity tracking builds on the work done in installation management, where the model has been used to show how a project will be constructed using LBS. However, construction activity tracking takes the resulting schedule to a more granular level to manage tasks that occur in the field, which can then be fed back to the project manager or superintendent in the field to address delays or to begin work early as construction progresses.

There are a number of tools that can take a BIM-derived construction schedule and break it into achievable tasks. Pull Plan is a SaaS tool that combines the lean methodology of planning with "sticky notes" and a cloud-based task notification system that alerts teams when they have tasks coming up or at task completion to select whether the activity has been completed on time (Figure 5.35). If a task is completed on time, then the activity is closed and the next task is activated. If the task requires more time, the user inputs how much more time is needed and the root cause for incompletion. The schedule is then automatically adjusted, and other users dependent on that task's completion will not be notified to start.

This tool is not necessarily for overall project scheduling; rather it is a way to take larger construction activities and break them down into smaller tasks that can then be measured. One of the more tedious aspects of a construction project is updating the project schedule. Because larger Primavera and Microsoft Project schedules become inaccurate at the first project delay, the use of in-field mobile-enabled feedback tools allows for capturing construction progress. Combined with in-field construction management tools such as BIM 360 Field and Navisworks, models can be color updated and overall daily progress can be captured in a much more meaningful way.

Figure 5.35 Pull Plan project

Field Issue Management

How quickly issues that occur during a construction project are resolved during a job can be the tipping point for a successful project. One of the biggest constraints is not in the identification of the issue at hand, but rather the speed and response to issues in a timely manner. Bulky construction management tools in the field are being replaced by nimble applications that allow project teams to use a smaller mobile-ready version of a larger application or a separate application altogether to get answers or direction for issues that occur on-site. So why do projects get held up in issue management and how can BIM help?

In a typical construction job, as issues pop up they are either determined to be the responsibility of the performing contractor to resolve or they are out of their scope and need to be clarified by a member of the design team. These issues are then sent out as RFIs or other formally issued clarification, with supporting data. This information may be photographs, narratives, or plan set revisions to show where more information is needed. Once created, these questions are then sent to the responsible party for answers. Sometimes these questions are automatically issued from project management software tools such as Procore, ProjectWise, or Prolog. From here, information then bounces back and forth to make sure the problem is understood until an answer is given or other direction is provided.

Many times, design staff don't fully understand the implications of untimely responses, and as a result, project delays occur and costs are incurred. This can get really messy in the form of change orders and other penalties. Because of the inefficiency in the traditional process, the construction industry has begun to question the tools used as well as the process itself. So where can BIM play a role?

Using Model Links   The use of model links to clarify issues can help as well as tools such as BIM 360 Field that use the 3D nature of the model to identify where the issue is and to more clearly illustrate what needs to be fixed. Two-dimensional plans aren't the most effective means of asking for information and as a result often require further clarification of the question being asked.

Collaborate on a Model via Web Meetings   A model can be used in the job trailer to walk a design team member through the issue at hand via web meetings such as Lync GoToMeeting, join.me, and Google Hangouts. Each of these tools allows the meeting participants to collaborate, share screens, take notes, and record the meeting.

Using In-field Videos   The use of in-field video is increasingly becoming a reality. where an on-site associate can walk up to the problem area and have a live video meeting with the designer to talk through the issue. This can be accomplished through tools such as Skype, FaceTime, and Google Hangouts for mobile.

Ultimately, BIM can be used for myriad purposes in the field and the industry continues to discover other uses for the model to help in building projects. In-field BIM can yield valuable information and is an effective way for teams to perform analysis and collaborate. As models become more and more accurate, the uses for effective use of BIM in the field will increase and become more relevant.

BIM and Safety

The topic of BIM and safety has been broadly discussed and analyzed in a number of forums, whitepapers, and industry presentations. The uses for BIM as it pertains to safer work include the following:

- Fall protection analysis (guardrails, scaffolding, etc.)
- 4D site logistics simulations showing on which days certain areas of a site are dangerous
- Safety training and site orientation
- Increased prefabrication efforts to minimize ladder times and unsafe installation positions

In many ways, BIM is just starting to scratch the surface of how it can be used for better safety. This is a broad area for improvement in construction with technology, and the potential is too large to ignore.

One of the consistent themes with BIM as it relates to safer projects is the automation of work on-site. As one of the superintendents I've worked with said, "No one can get hurt if no one is there." Simply reducing the amount of man hours on-site reduces the opportunity for safety incidents. BIM can play a valuable role in site automation and information gathering.

Drones and Photogrammetry for Safer Reality Capture

Black & Veatch Engineering is testing the use of drones equipped with GoPro cameras to capture existing site conditions, as shown in the following graphic. By using a combination of software from Autodesk—ReCap Pro, Project Memento, and Revit—they were able to capture existing exterior facility information.

The group began by navigating the aerial drone around the exterior of the building, taking multiple pictures, as shown in the following graphic. The ReCap software stitches the photos together and makes the data available for import into Revit. From there the facility can be modeled further from the information available in the scan.

The group intends to further leverage this technology by applying this technology in their Telecommunications business to reduce or eliminate trips for personnel up radio and telecommunications towers, which can often range from 400 to 1,000 feet high. This would allow information to be safely and accurately gathered, as well as high-definition (HD) photos to be taken without the unnecessary risk of sending personnel that high above the ground. The results are a full 3D Revit model and a mass model with HD photos layered on top, as shown here.

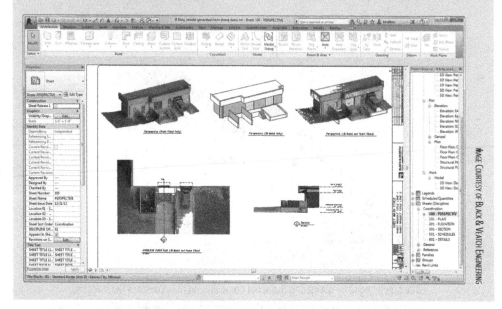

Although I have talked about some ways BIM can be used for safer construction projects, there are more opportunities that can be developed in this area—particularly in the use of mobile technologies and the use of models to layer in more intelligence about equipment, alert users of certain scheduled conditions, or automate evacuation instructions.

Producing Better Field Information

Why do you produce the information in the format you do in construction? Some of this information is created because there is a legal precedent and people feel comfortable with it; for others, it "has just always been done this way"; and some simply haven't found a better way of working. This section aims to have you—the construction industry—discover why you need the information you do, rethink the information you create and its format, and develop a more meaningful dialogue on improving the clarification stage of a construction project. There are plenty of books that talk about what an RFI is and why they are created, but I will suggest a different strategy for making information flow on a project site as well as better understand some of the barriers to disseminating data on a construction site.

Beginning with the End in Mind

The previous edition of this book explored new ways to issue RFIs, but this book is opening the dialogue to align to the larger context of questioning that is occurring within the construction industry. To start, you need to ask tough questions: "Why RFIs?" "Why construction documents?" "Why change orders?" Many of you may have heard industry presentations on zero RFI or zero change order projects. How did these happen? Are they coincidence or is there something larger at play going on that represents a better way of working together?

In Chapter 2, I showed you how to plan for BIM-enabled projects by beginning with the end in mind. This better enables teams to plan on getting to the desired end state, without focusing on traditional processes that may not add value. The same is true for construction. Although our industry is experiencing a technical renaissance, the ability to question processes and tools should extend into the traditional methods of creating, sharing, and answering project-related questions as well as collaboration overall. As Buckminster Fuller said, "You never change things by fighting the existing reality. To change something, build a new model that makes the existing model obsolete" (www.goodreads.com/author/quotes/165737.Buckminster_Fuller). This is the challenge for the construction industry, which traditionally has kept the tools used in circulation even though better and more accurate ones are available. So where do you start? To begin, I would like to challenge some of our thinking and constructs associated with how construction projects get built.

RFIs represent a failure in the current process. This may sound extreme, but when you think about it, it essentially means that some part of the contract documents were not clear or detailed enough to build from. Because information was not clear, the construction team must seek out additional information from the design team to determine and clarify intent. There are two issues at hand here. One is a topic that the AEC industry has been discussing at large for some time now—the relevancy of "design intent" drawings in today's modern world—and the other is a general lack of acknowledgment in how new technology should shift our industry practices.

Design intent was originally brought about as a means to formalize the idea that designers will never be able to create perfect drawings because of all of the moving parts, changes, and iterations a building design experiences. The unfortunate consequence of design-intent contract documents has resulted in designers who have withdrawn themselves from the interpretation process with the contractor and contractors who are left trying to play designer. The AEC industry could benefit from being more active participants in the design interpretation process, just as much as construction personnel should realize the massive benefits a project can gain by thinking through details in a model versus in the field. As Anshen+Allen Architects' Eric Lum says, "In today's environment, architects need to participate in new and innovative ways that facilitate a closer connection to the field condition" (www.aia.org/practicing/groups/kc/AIAB081947).

Relevancy in the design process is at stake, and designers and builders mutually need to find new ways of working just as they seek out new technologies. The concept of design intent drawings is increasingly becoming irrelevant because of what the BIM tools available today can accomplish. As discussed in Chapter 3, "How to Market and Win the Project," it is often easiest to bolt on additional tools, seeking out the "silver bullet" without challenging the process norms and looking at new technologies from a new perspective and asking what will this change in our processes. I believe the AEC industry at large still must further question the way we use models and shift beyond the 2D-only "contract deliverable" to a more technology-forward position that improves the process for all involved.

> *The first rule of any technology used in a business is that automation applied to an efficient operation will magnify the efficiency. The second is that automation applied to an inefficient operation will magnify the inefficiency.*

—Bill Gates

The AEC industry still needs to acknowledge the potential implications of BIM for what it is: an industry-changing technology, not a faster horse. In the beginning, the design industry looked at BIM as a way to create drawings faster, and many believe that this is the best use for this game-changing tool. In construction, while the investigation of "What if?" seemed to last a little longer, BIM was pigeon-holed as a tool for visualization and clash detection. In other words, it became a better tool for completing traditional workflows in document review and drawing coordination. I'm speaking in some generalities and there are always exceptions, but I believe it needs to be said that BIM should be a tool that shifts how we work. Our industry is hungry for *both* technology and process improvements. A big part of the work Virtual Builders is doing is to challenge and understand where existing tools, processes and behaviors can be improved on or completely altered for the better. I challenge anyone reading this book to be an active participant in being a part of this change. Ultimately, our industry will make the decision if BIM is simply a powerful tool that gets existing work done faster or if BIM becomes a central component of how we deliver the built environment.

So now that you recognize the challenge at hand, what are some of the ways BIM can shift the way you design and construct buildings? As this section heading states, it begins with the end in mind.

Leave the Box at Home

The following was related to us by David Miranda, director of maintenance, operations, and facilities at Tustin Unified School District:

"We often encounter construction managers and builders who market their ability to think outside of the box. Of course, many first look inside the box in search of quick solutions. Our preference is to contract with firms who recognize that there is no box! Those who don't make the shift to this mind-set will undoubtedly find themselves stifling their own creativity.

IMAGE COURTESY OF TUSTIN UNIFIED SCHOOL DISTRICT

"At Tustin Unified School District (TUSD), we are 'all in' so far as using technology to improve operational efficiencies. State-of-the-art technology is used in every classroom, and we've extended the TUSD Connect initiative into our Operations division. Building information modeling has been used as a construction visualization tool for many years, yet we are in search of builders who are forward thinkers and can take the model concept to the next level. School districts such as Tustin USD seek out firms who can use the model concept to provide integrated facilities information. We require comprehensive as-built virtual records that can be used by maintenance personnel to form a reliable basis for decisions during a building's life cycle from inception on. We believe that as an owner, continuing to push BIM and its use throughout the project life cycle encourages growth and innovation, while we as the customer get better collaboration and reduced risk."

What Information Do You Need to Build?

Interestingly, this question is being answered different ways by innovative teams who are pushing the boundaries of what it means to be a builder. Being cognizant of available, better technologies and understanding that the need for a builder doesn't necessarily have to be complex, you can develop elegant solutions that get to the core of what construction can be. For example, there are fewer limitations in file size or computing power for a designer's vision to be virtually represented to the smallest detail. BIM platforms—such as M-SIX's Veo, which significantly reduces model size yet retains intelligence, as well as the use of cloud environments or VDI/VDE infrastructure—make the limitations associated with the local desktop or laptop a thing of the past.

The ability to link components to information is now readily available in modeling and analysis platforms alike. Systems such as InterSpec's e-SPECS that connect specification data to model data provide the opportunity to get to necessary installation information. You've seen how models are being used for estimating and integrated scheduling with Navisworks and Vico. You've seen subcontractors begin to leverage the granularity in models to achieve tremendous efficiency gains (such as the Rex Moore case study). In-field activities are connecting to models via tools such as BIM 360 Glue and mobile applications. Vendors have opened up the APIs for modeling and analysis tools, further facilitating improvement through plug-in development, custom integrations, and other improvements to bridge information gaps.

So how will all of this available technology create change for design and construction teams to work together to virtually build a model so that the physical representation to follow is completely understood? This is the real challenge at hand and in many cases the elephant in the room. I'm not proposing that this will be easy to accomplish, or that there is a clearly defined roadmap to follow. Groups like Fiatech (Figure 5.36) are doing great work to explore the potential opportunities and how they can impact the project life cycle. You must challenge industry norms and have the dedication to question and understand what BIM could be.

Model Redlining Exercise

Using the model for construction is a worthwhile endeavor. In the following exercises you'll explore some of the ways models can be leveraged as an effective means of creating a more dynamic and meaningful dialogue.

The next exercise will focus on using Navisworks as an in-field construction administration tool. Navisworks provides construction staff with a means of defining what is to be built and checking against the existing conditions in the field. Best practices for handling issue coordination should be developed before construction begins. You should get input from the superintendent and field engineer, who are the ones most connected to the project and on the jobsite the majority of the time, and then determine workflows from there.

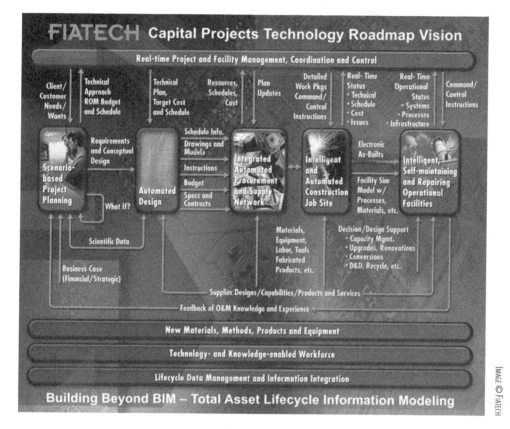

Figure 5.36 Fiatech's model roadmap vision

In Navisworks, these are the three tools used for commenting:

Comments   These are associated with clashes, saved views, selections sets, tasks, and animations.

Redlining   This is annotation that can be added over a viewpoint.

Redline Tags   These are tags used for recording issues during a review, and they combine the functionality of comments, redlines, and viewpoints.

Comments

For this example, you will use the Comments tool:

1. Open the Navisworks file `Construction-model.nwd`, and select a view you want to use.

2. Save the viewpoint by selecting Viewpoint ⇨ Saved Viewpoint ⇨ Save Viewpoint (Figure 5.37).

3. In the Saved Viewpoints tab, name the viewpoint **Corner Flashing**.

4. Right-click on the newly created viewpoint and select Add Comment (Figure 5.38).

5. In the Add Comment dialog box, type a comment or question you have, leave Status set to New, and click OK (Figure 5.39).

Figure 5.37 Saving a viewpoint

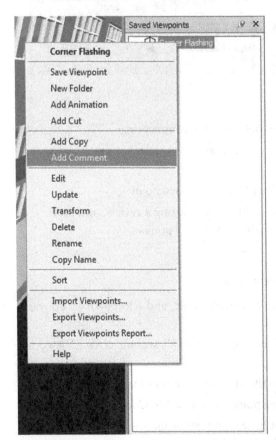

Figure 5.38 Adding a comment to the composite file

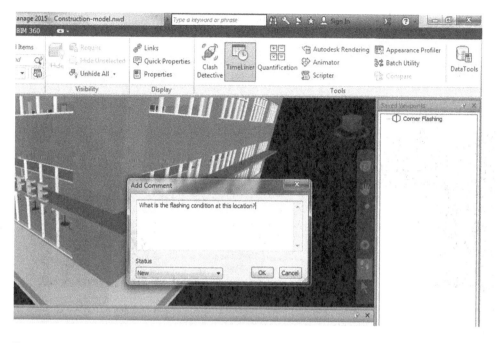

Figure 5.39 Saving the comment

You will notice that your new comment appears in the comment area, and it has tracked the author, the date, and time it was created, as well as its current status. Comments may also be found under the Review tab for reference. To edit the status of a comment:

1. Right-click the comment.

2. Select Edit Comment (Figure 5.40).

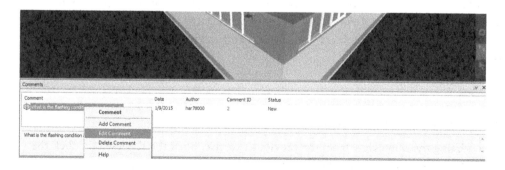

Figure 5.40 Editing an existing comment

This opens the Edit Comment dialog box.

3. Answer the question and change the status of the comment.

Figure 5.41 shows the status changing from New to Active.

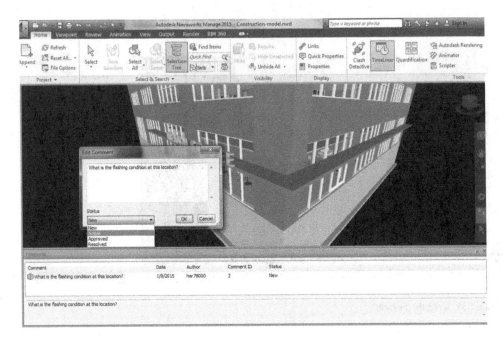

Figure 5.41 Comments on the model create related views and can be tracked and logged just like sheet comments.

Commenting alone is useful if you are looking to internally go through a model review or model issues in an organized manner. Usually comments are combined with redlining tools and other information to add more clarity to issues. Additionally, you can create folders using the Saved Viewpoints tab to better organize your comments and status of each revision. Just right-click in the window and select New Folder. Rename the folder to what you'd like and then drag and drop viewpoints into the folders. Next you'll add redlines to the model, using the redlining tool.

Redlining

Next you'll create a redline in Navisworks:

1. Using the same view from the previous exercise, open the Review tab, click the Draw tool, and choose the Cloud option (Figure 5.42).

    Other redline shape options include Ellipse, Freehand, Line, Line String, and Arrow.

2. With the Cloud option selected, make multiple clicks, circling the area you want to call out, to generate the cloud shape (Figure 5.43).

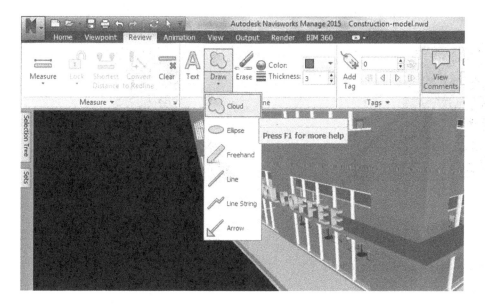

Figure 5.42 Opening the redlining tool

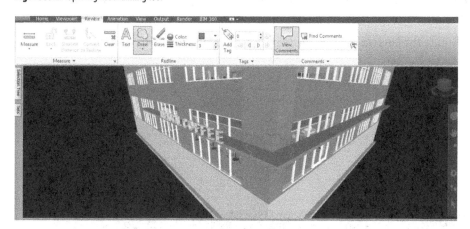

Figure 5.43 Creating the redline cloud shape

Note: You can edit the thickness before using the redline tools. In the Redline window, select the Thickness drop-down and adjust as needed. The default is 3″.

3. Once your shape is created, select the Text tool next to the Draw tool and click to close the cloud you just created.

4. In the Enter Redline Text field, type a comment and click OK (Figure 5.44).

Notice that the text entered shows up on the view but doesn't appear in the Comments window. This is because redlining is view specific. A more common revision method in Navisworks is the use of a redline tag.

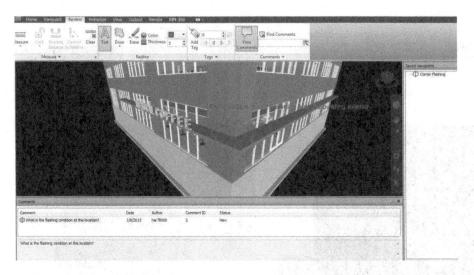

Figure 5.44 Adding a comment to the redline

Redline Tags

Redline tags are probably the most common means of marking up a Navisworks file. A tag automates view creation and lets you comment on anything in the model:

1. Rotate the current view in the Navisworks model to a new view, but don't save the viewpoint.
2. Select a model component.
3. Click Add Tag on the Review tab (Figure 5.45).

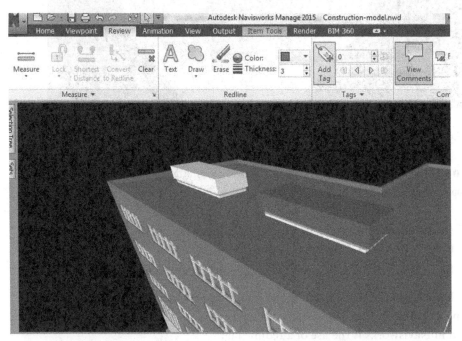

Figure 5.45 Adding a tag to model components

4. Click once where you want the tag leader to begin, and click again where you want the callout tag to end.

5. In the Add Comment dialog box, shown in Figure 5.46, insert a comment or question, and click OK.

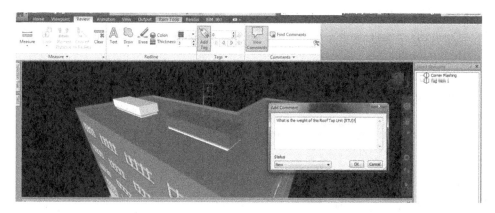

Figure 5.46 The Add Comment dialog box

You will now notice that there is a new comment in the Comments window as well as a new viewpoint called Tag View 1. You can rename it by right-clicking on the viewpoint and selecting Rename.

You can also create redline tags on clash reports by highlighting the clash in the report window, clicking the Tag tool, and associating a comment with it. If done correctly, it should have a red dot in the Comments window (Figure 5.47).

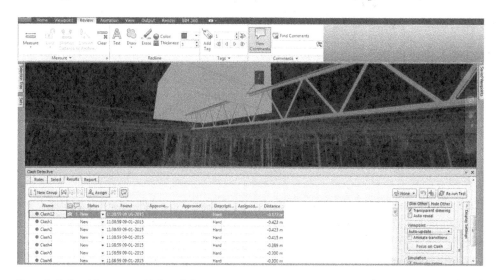

Figure 5.47 Tag with comment added to a clash item

You can search the tags by choosing Review ➪ View Comments and entering the search criteria for the tag you wish to find.

Video Embedding Exercise

Something to think about when using BIM is the augmented layer of reality or connectivity to the physical components that can create value. In this exercise, you'll see how to add links to a model to show conditions in a structure that are placed within the 3D constraints of the model:

1. Open the Construction-model.nwd file and navigate to a view where you want to put a link.

2. Select a model object using the Select tool from the Home tab, right-click on the object, and select Links ⇨ Add Link (Figure 5.48).

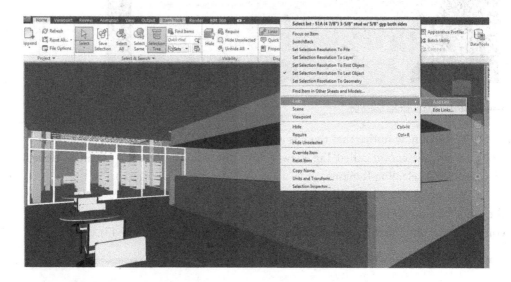

Figure 5.48 Setting up the link

In the Add Link dialog box, give your new link a name, copy the link or file location you want to add, and paste it in the Link To File Or URL field. In this example, I'm using a link to a YouTube video: https://www.youtube.com/watch?v=ydq21TWIYes.

Note: Keep in mind that file links will require access to the same network or file location that you have. So if you want to share a model with other team members, it is usually best to link to information that is available through the Internet. You don't want to link to information that requires access to a secure network or FTP. When possible, provide access to tools such as Box.com, Egnyte, Dropbox, or other web enabled sharing platforms that utilize hyperlinks to files.

3. Leave Category set to Hyperlink (Figure 5.49). If you want to specify where the link icon is placed on the object, click Add under Attachment Points and select where on the object you want the link placed.

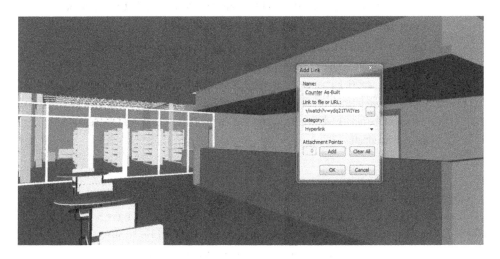

Figure 5.49 Defining the type of link

4. Click OK. You may need to click the Links icon on the Home tab toward the middle of the icon bar (Figure 5.50).

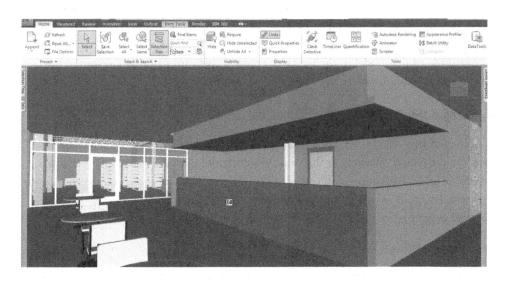

Figure 5.50 Links icon

Now if you navigate with your Select tool to where the link is located, it will act as a link to the file or URL you specified.

The Links Tool

Contractors use the Links tool for an array of purposes. Some use links to file repositories where videos, photos, cut sheets, or other information about that object are stored. In this method, the link acts as a miniature website for each object.

Other uses include using links as a way of sending and answering questions about a project. If the project is nearing completion, some owners like to put all of the operations and maintenance information about equipment within the model in the link locations. This is a powerful tool that allows accurate model data to extend to other information outside of the model, which increases the usefulness of the model as a single reference point of information.

BIM software continues to evolve in the field of BIM as a useful means of providing construction information. Many construction managers have a fundamental understanding of the software and practices for how BIM can be applied in the field. I've shown just a few of the ways a model can be leveraged during construction. Whether it's conflict detection and resolution, or schedule sequencing, models work well when used in the fluid process of design and construction, because conditions consistently change.

From design to construction and even into operations, a model's variance is often significant. Although BIM aims to limit the negative variation of issues and conflicts, construction is a complicated process, and it is unrealistic to expect no surprises when construction begins. This is where the value of a flexible process and continued use of the model in the field can create significant value as a resource to analyze and collaborate on to arrive at solid solutions.

The Virtual Job Trailer

So, how does BIM work in a job trailer, and what new technologies make the most sense? As the model migrates to the field, what technologies are needed to best support and enable its use? Although job trailers may never replace the corner conference room in the penthouse, there are some very useful tools that are being integrated on-site to assist site personnel in extending the life of a model and its effectiveness.

The Conference Room

To start, let's talk about the job trailer as a conference room (Figure 5.51). The primary difference between traditional projects and BIM-enabled jobs is that the models require computers or tablets to view and access. That said, a job trailer for a BIM project will need to have the ability to display the model and views generated from the model in the trailer.

Figure 5.51 Field coordination is where all the physical pieces come together.

There are different ways to address model viewing in the job trailer:

Touch-Screen LED TV   Large-screen touch displays are pricey and can range from $3,000 to $8,000 depending on size, but they offer the most versatility and intuitiveness when compared to other options.

LED or LCD Flat-Panel Screen   A less expensive option than the touch-screen display is a LED or LCD flat-panel screen. The cost of a flat-panel television can range from $500 to $5,000 off the shelf. These are probably the most common in BIM-enabled job trailers.

Large Computer Monitor   If a flat-panel display is not an option, the least expensive option is a larger computer monitor or projector. Monitors from 32″ to 60″ can be purchased anywhere from $300 to $3,000 and are as effective as an LCD or plasma display. However, they might lack versatility for some of the higher-definition or desired display-size functionality for multiple viewers.

Any of this viewing equipment can be reused on future projects and are worth the investment.

In a BIM-ready job trailer you should have Internet access and a wireless router. You also need a link to the display such as Apple TV, or you can use the share screen function on some smart TVs that links laptops with the flat-panel or monitor for ease of use when multiple laptops are being used in a meeting. This eliminates the need to swap out VGA and HDMI cords and is a great way to effectively communicate. This method is more environmentally conscious as well because of reduced printing costs, and comments generated during the meeting on a model may be combined into the same tool containing field comments for future direction and resolution without printing.

It is important that the room itself facilitates collaborative meetings. This includes the furniture, room size, room layout, and flexibility in configurations.

Think of how your project team will work, how many members will be attending the meetings, and what kind of work you will be doing, and then design the trailer accordingly. Remember it's not that expensive to customize a job office at the beginning, but it is more expensive to change it later once it's in place.

The Plans and Specifications Hub

The job trailer also functions as the plans and specifications hub. Increasingly, the industry is seeing a reduction in paper plans and specifications on-site. These are being replaced by servers linked to wireless networks, which allow field personnel to access the latest models and project information at any given time. Additionally, mobile applications with online and offline capabilities are making it easier for field staff to access project documentation in the field at the point of work. The key factors are:

- Mobility
- Internet access
- A single repository of project information that is continually updated

An example of this setup is the use of an online file storage tool such as Box.com or Egnyte, combined with a mobile-enabled editing tool such as GoodReader or Bluebeam Revu eXtreme. The ability to download the required information and sync data when the user is connected again makes the use of mobile technology in the field effective. According to George Elvin's *Integrated Practice in Architecture* (Wiley, 2007), the use of mobile devices in the field can reduce rework on a project by 66 percent when compared with more traditional methods.

The key element to this success is the remote access to network information, which allows field personnel to access the most current data anywhere on the jobsite. This is particularly effective on larger projects such as stadiums, warehouses, casinos, hotels, airports, bridges, and larger civic projects where even walking back and forth to the job trailer is an ineffective use of time, and lugging around large plan and spec sets isn't the easiest means of project coordination. This will be discussed in Chapter 6 in more detail.

The Jobsite Office as a Server

In addition to being the place where project stakeholders can view and access information, job trailers are increasingly becoming the point from which users sync to the latest data and achieve other tasks that may have been manual before. Here are some examples where the jobsite office acts as the hub for the push and pull of project information:

- The use of barcodes on hard hats to clock in and out workers
- Management of materials and equipment that are scanned with mobile devices to show status
- Safety reports that are automated and tied to models for commenting on where safety can be improved or where reported incidents occurred

There are tremendous efficiencies to be gained by automating repetitive functions such as reporting, quality control checks, safety audits, time keeping, and daily status updates. Many of these items can directly be tied to a model, which can then be further leveraged in the field if needed. Additionally, safety concerns related to carrying large plan sets and specifications on-site are reduced by making the information available on tablets and mobile devices. Of course, trips back and forth to the job trailer and the work can be reduced. Better information can be made available through the introduction of video content from field devices to project participants. Videos of actual conditions and observations make the process of issue management and in-field resolution go much more smoothly, though this technology is just being introduced as an acceptable means of requesting and clarifying in-field issues under current contract standards.

BIM use in the field gives construction staff a unique opportunity to develop and edit the information about a project from a single location and share that information with others. Instead of waiting until the end of the project to complete project as-built drawings, smart builders are updating the model as construction progresses to ensure that the model accurately reflects the conditions and work put in place for the customer. Then when construction is completed, they are sharing that model with project team members to show what was constructed. BIM is a much more usable tool in this capacity during construction as opposed to relying on redlined PDFs.

The Jobsite Office as a Communication Hub

When the industry first started using models in the field more than a decade ago, it was important to have videoconferencing capability. Meeting participants were less interested in seeing everyone's faces and more interested in seeing the model and visually following the discussion points. BIM is a great tool to use as a communication vehicle during construction because it's clear and everyone can interpret the 3D environment. Although it is still critical that project "humans" meet with other project "humans" to discuss and plan construction, the use of models as a tool to effectively represent the physical construction will become more prevalent.

Setting Up the Job Trailer

Correctly setting up a job trailer at the onset of a project is critical to successfully using BIM in the field. Although this might seem like somewhat of a menial task, this portion of the process requires a new way of looking at how to display and share information on a jobsite. As a best practice, look at setting up levels of tiers of equipment required that you can use to best enable your field team. For example, some firms use the approach of four types of jobsite office configurations: a small, medium, large, and extra-large configuration. Usually these configurations correspond to the size of the project, number of staff, and customer needs as they relate to technology.

Other construction management companies take the approach that each project is custom or that they wish to try out a new technology or configuration each time. This isn't the most effective strategy and requires support for each new project, but the custom-configured approach may very well be the right answer for certain customer bases.

When establishing your jobsite office standards, achieve a minimally viable level for each configuration. Factors to consider when deploying a BIM-enabled office are as follows:

- Internet connection (type and speed)
- Viewing configuration (panel type)
- Conference room requirements
- Mobility considerations (tablet and app support)
- Power demands (charging devices and screens)
- Flexibility
- Collaboration space

These factors represent a good start in developing a useful field technology configuration. However, there are many others depending on the type of work being done and the type of technology used. For instance, the use of laser scanners, drones, and robotic layout devices may very well shift the technology needed on-site to support their use. Often one of the biggest limitations to the use of BIM in the field is the lack of foresight by the contractor to plan and budget for its use during construction.

Summary

This chapter has covered using BIM during construction. All of these topics are specific to BIM processes, but they are not by any means the only coordination that can be accomplished on a jobsite. In a way, this chapter covers the basic "how to" of BIM during construction administration and shows some of the current solutions and shortcomings. From design to fabrication to construction, BIM should be the means by which a team references to and develops information. The construction industry is shifting. Though the pace of change may be slower than desired, the use of BIM on construction sites that create and share better information will only grow.

The BIM process during construction administration is continually being refined. Every day new case studies and methods of working rise to the surface and shape our industry to address what this exciting technology is capable of. The potential is great, yet it is important that tools be developed in the area of construction administration that address more accurate modeling and analysis strategies while having the courage to question and remove antiquated processes when needed.

BIM and Construction Administration

In 2009, when the first edition of this book was published, the majority of construction companies were faced with selecting or hiring a single BIM employee or developing a BIM department. We now realize that successful implementation cannot be restricted to an individual or department—it has to be elevated to companywide adoption.

To take it a step further, companies that will thrive in today's market are the ones that don't focus on one tool as "the answer" but that embrace a holistic view.

In this chapter:
The battle for BIM
Training field personnel
Document control
The real value of 4D
Developing BIM intuition
Small wins to big change

The Battle for BIM

The first, and most important, ability you can develop in a flat world is the ability to "learn how to learn"—to constantly absorb, and teach yourself, new ways of doing old things or new ways of doing new things. That is an ability every worker should cultivate in an age when parts or all of many jobs are constantly going to be exposed to digitization, automation, and outsourcing, and where new jobs, and whole new industries, will be churned up faster and faster.

Thomas L. Friedman, *The World Is Flat: A Brief History of the Twenty-first Century* (Farrar, Straus and Giroux, 2005, 2006, 2007) Release 3.0

A few resounding themes are found throughout this book, but there are two in particular that we'd like to focus on in this chapter: BIM is a tool and BIM is only as good as the hands that are using it. The people who can make the most impact with BIM in construction are the field personnel or the "administrators" of the construction (superintendent, project manager, project engineer, and so forth). These are the individuals who are on the front lines every day, interacting with the owner and dedicated to the quality of their work and the safety of their employees. They are the ones with mud on their boots, who are at the job before the sun comes up and who are the last ones to leave. They have a passion for what they do and enjoy the euphoric smell of freshly poured concrete and mud on the walls. It's these individuals who come together as a team to execute the dream of the design, and it's these teams that must have the tools they need to execute the dreams of the future. The "battle for BIM" is the internal conflict construction companies are facing with getting the tools in the right hands.

The book *Relentless Innovation: What Works, What Doesn't—And What That Means for Your Business* (McGraw-Hill, 2011) by Jeffrey Phillips does a great job of explaining the challenge construction companies have to confront in order to contend in the new market. In summary, *whenever an innovation threatens a company's foundational practices, it will be met with resistance.* Construction companies are not built on innovation; they're built on operations. For some construction companies, BIM is still viewed as an innovation and not an operational practice required for administering construction.

BIM-enabled individuals make up a minority of the operational workforce, and their contribution to the bottom line is still being evaluated by the majority who are generating most of the income for the company. The majority work under a policy and procedures manual that has been formulated over the years by the brightest minds in the company. Construction companies invest thousands or millions of dollars into training the majority on ways to execute these procedures. This manual is the benchmark for evaluating performance against your peers and ultimately determines promotions, salaries, and bonuses. It is performance based, clearly defined, and a measurable standard to work toward—if you want to become a superintendent complete steps 1, 2, 3 . . . . If you want to be a project manager, complete steps 1, 2, 3 . . . .

BIM can be perceived as relatively new compared to traditional methods, isn't universally accepted, has uncertainty to its value (especially when the tool is in the wrong hands), requires assumptions, isn't quite measurable, and doesn't necessarily determine the majority's promotions or bonuses. These perceptions make it difficult for the majority to understand the promise of BIM and sacrifice time away from executing the day-to-day operations known to determine promotions and bonuses. Given these circumstances, can you blame the majority for not adopting BIM? No.

The BIM minority, on the other hand, dwells in the promise of BIM. They use the software every day and see all the benefits that have been discussed throughout this book. In their minds, BIM is not necessarily a new tool, because they recognize we're in the early to late majority phase and not the early adopter phase, so BIM has proven to be the new "business as usual." With that understanding, they become easily frustrated and deflated when the value they feel they're contributing isn't recognized by the traditional majority. These internal divided visions from the *traditional majority* and the *BIM minority* have created a battle for BIM and inconsistency in executing BIM effectively.

> We've seen large contractors that we thought were going to do very well, not
> do very well. We've seen small to medium size...just kill it on the BIM side.

Source: Phone conversation with Craig Dubler, virtual facilities engineer at Penn State University and project manager for the Penn State BIM Execution Planning Guide

Note: This resistance to change is even harder for the larger, more established companies. It makes it that much more critical for these companies to focus on their strategy for companywide adoption.

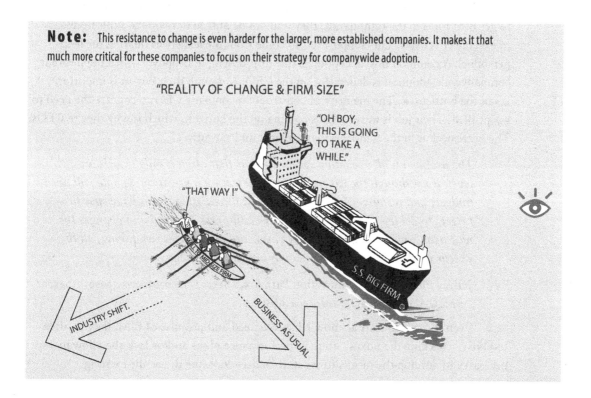

"REALITY OF CHANGE & FIRM SIZE"

This inconsistency creates a tremendous threat to companies unwilling to disrupt the traditional process, because though the internal value is being argued about, the external value is being revealed. Owners and subcontractors are starting to develop matrices based on contractor BIM performance, which means the value of BIM is starting to be quantified. Since 2007, the people who were most affected by a contractor's careless distribution of BIM tools to inexperienced professionals were these two groups. Now both "clients" are analyzing the results and finding ways to deal with these issues:

Owners Owners have started using standards to weed out the "Hollywood BIM" and the teams that oversell and underdeliver. The implementation of these standards will continue to grow and become more robust as owners analyze the successes and failures of projects that have used BIM.

Subcontractors Subcontractors have a different media to sort out the "Hollywood BIM." It's called "fee." They are starting to raise their fee depending on how well contractors administer BIM, because they lose money on detailing, material, and field labor when coordination is done poorly.

What this means is that contractors might not even be able to compete for work, based solely on qualifications from the owner. It also means that if they make it through the qualifications, they may not be able to compete due to estimates from the subcontractors. Executing BIM effectively is no longer an option—it's a necessity.

The challenge is that as long as companies reward the traditional practices and fail to get BIM tools in the right hands, they will not be able to develop the skills necessary to execute effectively. This support has to come from the leaders evaluating the field personnel. A common misconception is the "top-down philosophy," which implies that companywide adoption is dependent on the CEO. Although their buy-in is important, it is not the bottleneck. The majority of construction company CEOs recognize the need to adopt BIM. Their job is to see what's coming on the horizon, which is why they're CEOs. The bottleneck is in the mid-level management and executives.

> *The problem with executive sponsorship is that often it only travels one level. Even though the executives are advocating innovation, people still need budgets and resources for innovation and the existing teams have specific targets to achieve. Unless the executive team gets involved and changes the way people work on a day to day basis, and encourages risk taking, all the executive sponsorship is just a communication strategy.*
>
> Jeffrey Phillips, "The Innovation Paradox," http://innovateonpurpose.blogspot.com/2008/10/innovation-paradox.html

Even if the field staff acknowledges the need and promise of BIM, they're often too busy "checking the boxes" on their performance plans and/or lack the allowances necessary to develop the means to see it. If leaders continue to use the existing procedures to track their staff's performance against their peers, it will be difficult for the majority to believe they should be transitioning to this new process. Field personnel must be given the resources to bridge the knowledge gap, but also have a willingness to learn how to get across it (Figure 6.1). A great interdependent relationship can be created between the BIM minority and the traditional majority, because the minority will struggle without the majority's support. At the same time, the majority needs the BIM minority so they can synthesize their practical knowledge with innovative technology in order to leverage it and adapt to this new industry.

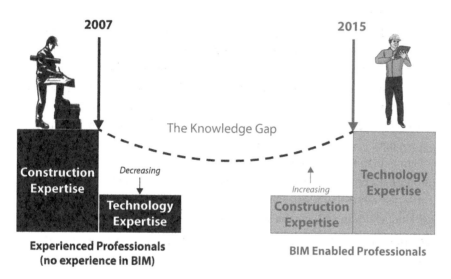

Figure 6.1 The knowledge gap

Training Field Personnel

You wouldn't hand a man a tool without training him how to use it; otherwise, he's likely to hurt himself or someone else. The same rule applies to BIM. The tools are getting more intuitive, but there's still a big learning curve. Field personnel training should be relevant, focused, and to the point. Field staff will typically not need to know how to create linked estimates, create sequencing animations, or draft BIM contracts. However, they will need to have a basic understanding of the contract language and the ability to leverage the model for various tasks: validating installation, tracking issues, managing changes and punch lists, commissioning, and closing out projects. In addition, they might learn how to generate a logistics plan, do virtual mock-ups, or run a clash detection in the field. Overall, it is typically the way the company is structured and how it wants to run its operations that will determine how much, and to what level of detail, the field personnel on a BIM project are expected to accomplish and be trained.

The ideal timing for training any staff is when they've completed a course in BIM and are immediately getting placed on a project to which it applies. When this method of training is not possible, it is best to schedule BIM training prior to the BIM project beginning. An improper assumption made too often in the industry is that once an associate is trained, that associate will retain all the information from the course whenever the next BIM project comes up. Instead, all too often the selected individuals will have intermediary projects in which they fall back into accomplishing tasks through the traditional process. Then when a BIM project does start, they are met with a certain level of frustration because they have forgotten the BIM training they received or the tools have changed.

Generally speaking, training field personnel is somewhat different from training office personnel, and the levels of education and understanding of software can vary widely. The challenge in providing training for new BIM software and new processes to field personnel is gauging the level of associates. If possible, you should group new users with associates who have already completed training or have prior experience and can support the other. Many times this methodology is most effective when a senior associate has a younger associate who can help with the new processes and programs alongside the senior staff member. In turn, the senior associate can provide experience and insight into in-field project management to assist the other associate. The BIM department can assist remotely or visit jobsites to answer questions and provide support. Either way, field associates must have the following to successfully leverage BIM:

- An in-depth understanding of the existing vs. new processes as defined by their organization
- A basic understanding of the software
- An understanding of how BIM software can make the project more coordinated
- An ability to evaluate what will save time and what tasks will take more time
- A support system (either in-house or external)

Field personnel typically retain the most knowledge with in-the-field training. This method connects the dots from the software and helps them apply what they've learned to tasks they need to accomplish (Figure 6.2). To implement a solid foundation for training field personnel, begin with an introductory course into modeling software such as Autodesk Revit or Graphisoft's ArchiCAD. You can often find courses through your local software vendor, training consultants, or online providers; these courses are useful for outlining the basic functionality of the applications and foundational understanding of the "I" of BIM. This should then be followed by a small group or one-on-one training, internally, to describe how these basic skills can be applied to day-to-day tasks and company processes. In describing the applicability, the training will lead to incrementally more advanced levels in two other areas: coordination and mobile field applications.

Figure 6.2 Superintendent using BIM in the field to coordinate construction—applying learned technology to the field

Note: Many times the first group to complete a high level of training becomes a resource to the company so that there is no need to engage outside resources for training afterwards.

Training Goals for Basic Skills

Make sure that the course outlines basic modeling skills, such as how to create walls, doors, ceilings, and so on. Also verify that the course outlines how to edit these components. Remember that the purpose of training the staff is to gain an understanding of how BIM is different from CAD or paper drawings, to learn how they can drill down to find needed information within the model, and to understand how they can use the model as a functioning piece of building information in the field. Additionally, this training provides the team with an understanding of *time*. How much time does it take to update model content? If the project team doesn't have a basic grasp of the time required to produce model content, then they'll have no way of scheduling coordination or managing change orders, two critical aspects of a successful job (schedule and cost).

Advanced Training Goals for Model Creation

More advanced training in model creation software is critical to self-perform construction with BIM, managing the evolution of the fabrication BIM, and delivering a *record model* to the owner. It is important for the field personnel receiving training to gain an in-depth understanding of the modeling software and an advanced level of training in how to integrate and know what to expect from other team members.

For example, if the contractor is pouring the concrete for basement walls on a project, they will potentially need information from MEP subcontractors for sleeve and penetration locations, as well as embed information from the steel and/or miscellaneous metals subcontractor. The training should clarify:

- What information is required for the construction BIM (means and methods)
- Where the required information is located (toolbelt)
- Whether or not the information is useful and accurate
- How to import the information and/or create the missing information to complete the construction BIM

The current reality is that too often fragmented, unreferenced, and inaccurate data is distributed between the construction team and then handed over to field staff for quality control and layout. Or that data is given to the owner to be used for the maintenance of the facility.

If the owner is requesting a record model, the creation of the record BIM from the construction BIM must be completed as accurately as possible to adhere to the standards requested by the owner and facility manager. If it is not the intent of a contractor to train their associates to such a high level of modeling skill, a number of strategies for delivering a record model might influence the decision to train field personnel in-depth:

Architect-Controlled Record Model   One option is for the architect to maintain a record model and update the model as the field personnel sends the AE team updated information. The format for this information can be typed documents, sketches, faxes, or models. The architect then reviews all documents and updates the record model. The model provided by the staff is often the more useful one because altered sections can be overlaid with model linking so the architect can see changes to the building and update the model. The advantages to this are that the architect (who might be more familiar with the modeling software) is able to maintain and track an accurate representation of the facility as the project nears completion. The disadvantage to this type of process is that often an architect will not have enough construction administration time budgeted to have the staff update the model. Although this should typically be a change in BIM process, sometimes this is not a feasible option.

Field-Controlled Record Model   The second methodology for a model to be updated in the field is for the staff to update the model and send the model to the architect for approval throughout the construction process. Doing so requires a higher level of training, but it is a good way of increasing efficiencies while still maintaining the accuracy of the BIM file that the field personnel are using to construct a building. Although this method doesn't require as much time from the architect, there still may be some push-back on the amount of time budgeted during construction administration to review these models.

Third Party–Controlled Record Model   Other times a separate construction manager, owner's representative, or virtual construction consultant might be present in a project and have the ability to edit the construction model and provide a record BIM.

For the most part, field personnel use modeling software to verify installation, check dimensions, locate fabricated components, update the record BIM, and visually communicate to subcontractors a complex or 3D reference. Other software can provide dimensional references; however, the tools, snapping, and interface are typically accurate in the modeling software. It is best to outline at the beginning of a project who will be responsible for using and editing the construction BIM through this software, which will eventually become the record BIM.

Training Courses for Additional Uses

After they've mastered the basic skills, field personnel should be trained to leverage BIM files for additional uses. This training should teach associates how to maneuver in coordination software and combine trade models for analysis. More specifically, the training should show field teams how to tag issues within the model, manage clash detections, integrate laser scanning technology for QA/QC, and translate information across BIM platforms (desktop or cloud-based). An example of information translation will be shown later in this chapter.

Once field staff are trained and have the tools they need, the promise of BIM will begin to reveal itself. The traditional majority will find new and innovative ways to use these tools and beyond what the BIM minority thought was possible (see the sidebar "Installation Verification Using Laser Scanning: The iMRI Installation at Kaiser Oakland"). The ultimate goal will be to have these BIM-enabled individuals participate in preconstruction, where they can apply their practical knowledge with this new technology, and continue to leverage the tools into the construction phase. This will eliminate any information loss between the two phases, help solve issues when they are most cost-effective, create high-performing buildings, and allow field staff to lock up the gates before the sun goes down.

Installation Verification Using Laser Scanning: The iMRI Installation at Kaiser Oakland

Intraoperative imaging involves the installation and utilization of scanning equipment typically located in a hospital's radiology department—such as magnetic imaging resonance (MRI), computed tomography (CT), or angiography—within the operating procedure rooms. This allows surgeons access to real-time or near real-time imaging

Continues

Installation Verification Using Laser Scanning *(Continued)*

during surgery where they can adapt surgical planning and accomplish surgical goals with improved precision and outcomes (intraoperative MRI [iMRI] shown in the following graphic).

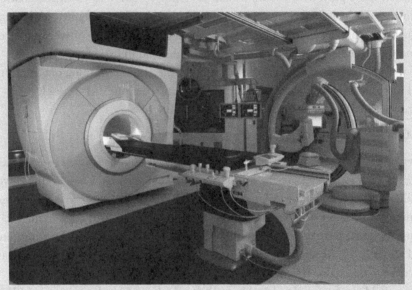

© IMRIS, INC. 2015

Intraoperative imaging was originally developed for the treatment of brain tumors. Imaging helps physicians better identify and differentiate diseased tissue and increases the likelihood that tumors would be completely and successfully treated.

For example, when using an MRI, the neurosurgeon must develop a plan related to navigating through the skull and brain for resecting tumor tissue based on images taken days before the operation. The surgeon will conservatively remove as much tumor as possible, taking care not to affect functional areas, and then send the patient for recovery. When possible, the patient will be sent for a follow-up MRI; if there is remaining tumor found on the scan, the physician may have to perform a second operation. With iMRI, the physician has the benefit of scanning and reviewing a patient's images multiple times during the first operation. In addition, the iMRI images help the surgeon adjust for "brain shift," where the tissue and skull changes shape during the procedure.

What Is the IMRIS Solution?

High-field iMRI operating suites are configured in one of two ways. Both require that the MRI scanner be stored in an adjacent room so that the clinical staff can use standard OR instruments during the surgery. One configuration requires that the patient be transported from the operating room to the magnet in an adjacent room to obtain an image. The second configuration (only offered by IMRIS, Inc.) moves the MRI scanner to the patient via ceiling-mounted rails to obtain the image (as shown in the following graphic). The latter approach has the advantage of not moving the patient from the operating theater during the surgery and enhances workflow and safety in terms of airway control, monitoring, and head fixation. This was the approach used on the iMRI installation at Kaiser Oakland.

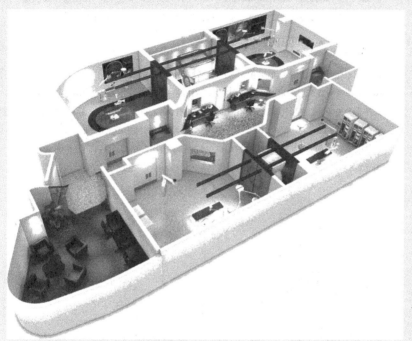

© IMRIS, INC. 2015

iMRI Support Steel Requirements

Imagine the sensitivities and tight structural tolerances required for smooth travel of an MRI system weighing nine and a half tons. Compound that with seismic regulations in California for earthquakes and the stringent requirements of OSHPD (Office of Statewide Health Planning and Development) for building in an acute care facility. This combination of requirements was met head on by McCarthy Building Companies' Kaiser Oakland project team, who installed the first IMRIS suite monitored by OSHPD and exceeded the following structural requirements:

- Structural beams must be leveled, parallel, and straight over the entire span of rail travel at +/– 1/8".

- Measurements must be taken every 21" to the whole length of the beam.

- Max deflection must be less than 0.156" between 7'-0" sections of magnet travel and 0.108" between parallel structural support beams.

Continues

Installation Verification Using Laser Scanning *(Continued)*

Laser Scanning for In-Field Verification

Due to the precise steel verification required by IMRIS and McCarthy's knowledge and expertise on surveying and laser scanning, McCarthy decided to use laser scanning in lieu of traditional surveying methods to analyze the support beams and concrete floors in the IMRIS suite.

Traditional Method

The steel trade partner constantly monitored the steel installation with a survey transit that was based on an established benchmark that McCarthy set. When the steel installation was complete, McCarthy had the steel contractor verify that they met the vendor's strict requirements by providing a survey at the same exact points that the vendor was going to check. Accuracy of the survey transit is ±1/8″ max. This traditional survey method took the steel contractor six hours to complete (see the following graphic).

IMAGE COURTESY OF MCCARTHY BUILDING COMPANIES, INC.

Laser Scanning Method

McCarthy's field personnel concurrently used a Faro Focus 3D laser scanner to conduct a scan of the support steel in order to produce a 3D point cloud. The total field time for the collection of data was less than 30 minutes. The data was then processed using Scene (Faro's laser scanning software) to register field scans together and then cleaned for analysis using Trimble RealWorks (shown in the following graphic).

Afterward, McCarthy used a combination of Autodesk products (AutoCAD, Civil 3D, Revit, Navisworks Manage) to accurately analyze the point cloud data according to specifications given by IMRIS and produced documents containing the survey results within 2 hours. Additionally, they did periodic laser scans of the floor to do quality control and levelness studies (shown in the following graphic).

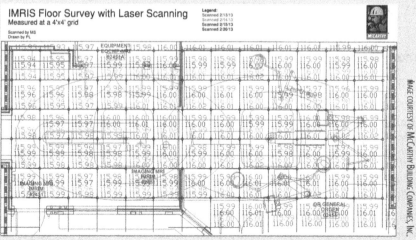

Continues

Installation Verification Using Laser Scanning *(Continued)*

This innovative method significantly reduced the time spent in the field to 1/8th of the time it takes to survey the support steel traditionally. This reduced not only conflicts with other trades who had competing objectives, but it also provided a higher level of accuracy. Additionally, because the analysis of information via point clouds were automated, the results were less prone to human error. There was never a need to return to the jobsite to recheck problem areas—the point cloud was so rich with information that it's akin to having the actual steel members in your office.

Setting New Industry Standards

The wealth of information that the 3D laser scan captured helped McCarthy set higher standards of quality for IMRIS to implement worldwide. McCarthy identified and mitigated potential cost and schedule impacts that arise with traditional surveying methods, and they demonstrated with scientific and statistical evidence that laser scanning performs at a standard of accuracy that is not easily attainable by other methods. IMRIS came back to reconfirm the results themselves, as they had experienced negative outcomes with other hospitals having issues with IMRIS steel—despite inspection results meeting their own tolerance requirements for installation.

"IMRIS completed its own survey of the steel McCarthy has installed for its Intraoperative MRI machine and reports the alignment is within a millimeter of optimum. This is the best steel installation IMRIS has seen in the last five years of worldwide equipment placements." —Customer testimonial from Jim Kautz Sr., project manager at Kaiser Permanente National Facilities Services

COURTESY OF © IMRIS, INC., KAISER PERMANENTE®, AND MCCARTHY BUILDING COMPANIES, INC.

Document Control

During the construction phase of a project, the flow of 2D information in a project increases. Specifically, addenda, supplemental drawing information, RFIs, and submittals, to name a few types of information, start to be produced more rapidly than before. The success of the project is usually a direct result of how well this rapid influx of information is managed, tracked, and distributed. Many times this involves the use of a gatekeeper, a single person or team responsible for managing the information coming in and then distributing it to the rest of the team. The role of gatekeeper may be a project manager, project engineer, BIM manager, or other personnel. The information distribution's success depends on the ability of the gatekeeper to communicate to all parties, to make sure that the correct data is distributed to the relevant people, and to manage the documentation.

Traditionally speaking, document control on the jobsite has been paper based and organized using massive filing cabinets, three-ring binders, and sticks of drawing sets separated by trade: Architect, Structural, Mechanical, and so forth (Figure 6.3). It's the gatekeeper's responsibility to make sure the filing cabinets, binders, and drawings are kept up-to-date with the latest information to ensure no one is building off old or outdated information. For example, once an RFI is returned from the design team, the gatekeeper is responsible for clouding and labeling the affected area(s) on the floor plans, elevations, and/or details in the sticks of as-built drawings, so the project team is aware of a change. In addition, they file the answered RFI in a binder or filing cabinet so that any team member can review the drawings, easily recognize clouded areas, and locate the RFIs that correspond to their work. Periodically, the design team will submit updated drawings to reflect the changes noted in the RFIs. These drawings are then "slip-sheeted" into the as-built sets, a manual process where the gatekeeper inserts the new sheets into the as-built set and archives the old ones. They are then responsible for transferring all clouds from the old sheets onto the new ones, which can be a long and arduous process.

Figure 6.3 Sticks of as-built drawings

Fortunately, the construction industry is beginning to embrace paperless workflows for the submittal, RFI, and drawing review process, which creates efficiencies in organizing, updating, and navigating documents, as well as improving the quality of life for the gatekeeper. The following exercise will walk you through how to set up a hyperlinked and organized digital plan room using Bluebeam Revu eXtreme, which automates document control on the jobsite.

Creating a Digital Plan Room with Bluebeam Revu eXtreme

Bluebeam Revu (www.bluebeam.com/us/bluebeam-difference/) is a powerful PDF viewer, creator, markup, and collaboration tool with plug-ins for Microsoft Office, AutoCAD, Revit, Navisworks Manage, Navisworks Simulate, SketchUp Pro, and SOLIDWORKS. It's currently available in three different packages with varying costs and capabilities: Standard, CAD, and eXtreme. Although CAD and eXtreme allow you to export 3D PDFs directly out of the modeling platforms for analysis or RFIs, this exercise will focus on 2D document control. For this exercise, you'll need Bluebeam Revu eXtreme, because of its OCR (optical character recognition), AutoMark 2.0, and Batch Link capabilities.

OCR   Allows you to create editable and searchable documents from scans or PDFs

AutoMark 2.0   Automates page labeling

Batch Link   Automates the hyperlinking of documents

Organizing Your Toolbelt

Creating a *folder structure* and a *naming convention* are standard processes that must be followed in both document control and model control. For this exercise, create three separate folders to represent the sticks of drawings, filing cabinets, and three-ring binders:

- Drawings
- Submittals
- RFIs

Creating Page Labels

With design-bid-build, as well as integrated delivery methods, design teams will submit their drawings at various milestones as a large combined PDF (Figure 6.4). When you open these PDFs in Bluebeam, you'll notice that none of the pages have a reference other than their sequential order. In order for the gatekeeper to create hyperlinks and virtually slip-sheet, the pages need to be separated and labeled, as follows:

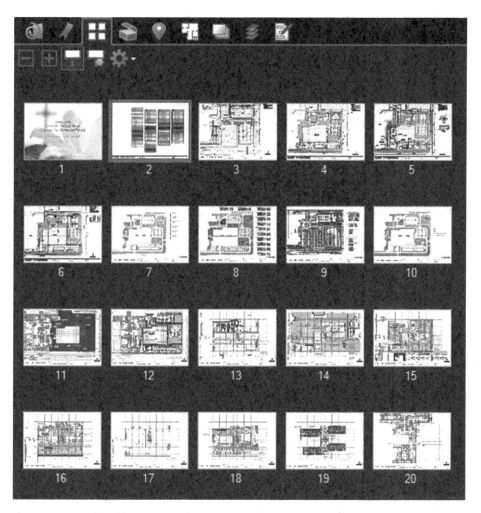

Figure 6.4 Combined set of drawings with no labels

Note: If the PDF doesn't contain searchable text, you'll need to run the OCR under the Document tab or press Ctrl+Shift+O, as shown in the following graphic, before continuing.

1. Open the AutoMark 2.0 tool, select Page Region, and click Select. This tool is located just below the thumbnail view icon and its quick properties say "Create Page Labels" (Figure 6.5).

Figure 6.5 AutoMark 2.0 icon

2. Zoom in on any sheet where the page label is located and draw a box to fully encompass the label.

3. In the AutoMark window, verify the label is correct in the preview and click OK.

4. In the next window, make sure all pages are included in the page range and click OK.

These steps will automate the page labeling process. Notice how all the sheets now are named by their label (Figure 6.6).

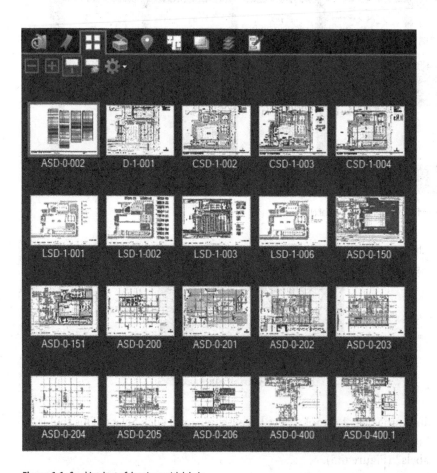

Figure 6.6 Combined set of drawings with labels

Extracting Files by Label

The next step is to separate the sheets and create your virtual sticks of drawings. Separating the documents allows the gatekeeper to automate the slip-sheeting process:

1. In the Drawings folder on your computer, create subfolders by discipline (Architectural, Structural, Mechanical, and so forth) to help organize your sticks.
2. Use the thumbnail preview to select multiple sheets.
3. Under the Document tab, select Pages ➪ Extract Pages, enter the settings shown in Figure 6.7, and click OK.

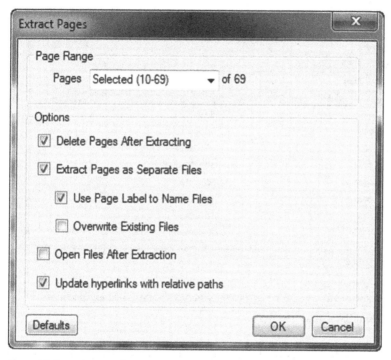

Figure 6.7 Extract Pages window

4. Navigate to the appropriate discipline's folder and click Select Folder.
5. Repeat steps 2–4 until all of your sheets are in their virtual sticks.

Hyperlinking All Documents

Once all the virtual sticks are in their designated folders, you'll want to hyperlink them together so you can easily navigate between floor plans, details, and elevations:

1. From the File tab select Batch ➪ Link ➪ New.
2. Select Add Folders and navigate to the Drawings folder on your computer (which now contains all of the virtual sticks), select the folders you wish to hyperlink together, and click Next.

3. In the Batch Link window, select File Name as the search term.

4. Verify that the Scan Result and Search Term match under the Scan Preview and then click Generate (Figure 6.8).

5. Under Link Options, you can change the color of the overlaid hyperlinks. After adjusting to your preferred color, select Run.

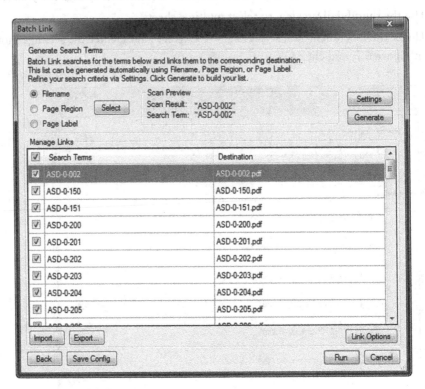

Figure 6.8 Batch Link window

These steps will automatically link all the floor plans, details, and elevations in seconds. Navigate to any sheet and check it out (Figure 6.9)!

Hyperlinking RFIs

You can now use these hyperlinked plans to manage the RFI process with the cloud and text tools contained within Bluebeam Revu:

1. Navigate to a floor plan.

2. To create a cloud, use the shortcut command by clicking "C" on your keyboard and draw a box around any object on the drawing.

3. To create a text box, use the shortcut command by clicking "T" on your keyboard and draw a box near the cloud; then write an RFI description (Figure 6.10).

ASD-1-100	FLOOR PLAN - LEVEL A
ASD-1-101	FLOOR PLAN - LEVEL 1
ASD-1-102	FLOOR PLAN - LEVEL 2
ASD-1-103	FLOOR PLAN - LEVEL 3
ASD-1-104	FLOOR PLAN - LEVEL 4
ASD-1-105	FLOOR PLAN - LEVEL 5
ASD-1-106	FLOOR PLAN - LEVEL 6
ASD-1-107	FLOOR PLAN - LEVEL 7
ASD-1-108	FLOOR PLAN - LEVEL 8
ASD-1-109	FLOOR PLAN - LEVEL 9
ASD-1-110	FLOOR PLAN - LEVEL 10
ASD-1-111	FLOOR PLAN - LEVEL 11
ASD-1-112	FLOOR PLAN - LEVEL 12
ASD-1-113	FLOOR PLAN - LEVEL 13

Figure 6.9 Sheet index with highlighted links

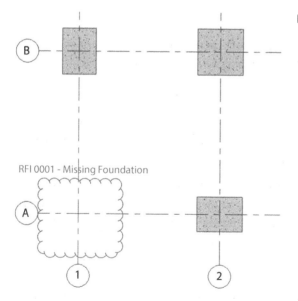

Figure 6.10 RFI cloud

RFI 0001 - Missing Foundation

4. Right-click the RFI text and select Edit Action. Do not group the cloud and RFI text—doing so would create a large hyperlink window, which would be problematic when there are multiple RFI clouds on a sheet.

5. Set the Action to Open and browse to the RFI or submittal related to the issue or item (Figure 6.11).

6. Select the Use Relative Paths check box and click OK. This will close the Action window.

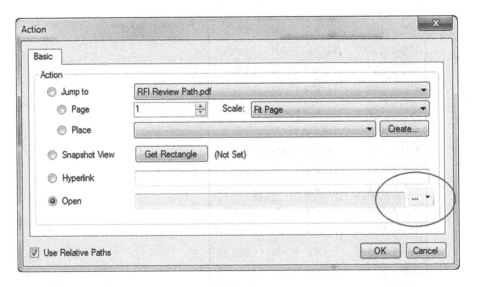

Figure 6.11 The Action dialog box

 Note: Relative paths allow links to work as long as files remain in the same folder location with respect to each other. Be sure to check Use Relative Paths if you plan to share the digital plan room.

7. On the Document tab, select Flatten and select Allow Markup Recovery (Unflatten); then click Flatten.

Congratulations! You've just posted your first digital RFI and hyperlinked it to the actual document. You can use this same procedure to hyperlink submittals, specifications, models, operation and maintenance manuals, websites, and so forth. Hopefully, you'll never look at three-ring binders again.

Digital Slip-Sheeting

The final step is knowing how to slip-sheet when an updated PDF is delivered from the design team or subcontractor:

1. Save a copy of the outdated sheet in an archive folder.

2. On the Document tab, select the Flatten drop-down and choose Unflatten All Markups on the outdated sheet so they will transfer to the new one.

3. On the Document tab, select Pages ⇨ Replace Pages.

4. Navigate and open the new sheet.

5. Verify that you are replacing the right sheet and select the Replace Page Content Only check box. Click OK (Figure 6.12).

6. Flatten the transferred markups on the new sheet.

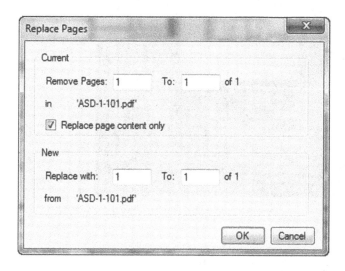

Figure 6.12 Replace Pages dialog box

That's it! This will automatically transfer all the posted RFIs and hyperlinks to the new document.

Note: Typically on projects you'll be slip-sheeting more than one document. In this situation it will be more beneficial to use Bluebeam's Batch Slip Sheet function to automate the archiving and replacing of multiple sheets. Go to www.bluebeam.com/us/bluebeam-university/whats-new/batch-slip-sheet.asp to learn more about Batch Slip Sheet.

What used to take days to accomplish now takes minutes. The amount of time saved for the gatekeeper is significant but pales in comparison to the value this process brings to the entire project team, including owners, architects, engineers, and subcontractors. This process mitigates the risk associated with working off old documents. It makes navigating through drawings, submittals, and RFIs effortless by using hyperlinks. The entire plan room can now be uploaded to the cloud or downloaded to a flash drive to share on mobile devices and kiosks in the field (Figure 6.13), as opposed to lugging around full-sized sets (Figure 6.14). Bluebeam Revu eXtreme is not the only tool that allows for this automation, but with a current price point of $349, it makes a very strong addition to the field staff's toolbelt.

IMAGE COURTESY OF McCARTHY BUILDING COMPANIES, INC.

Figure 6.13 Mobile kiosks in the field

IMAGE COURTESY OF McCARTHY BUILDING COMPANIES, INC.

Figure 6.14 Superintendents looking at drawings in the field

The Real Value of 4D

Model-based scheduling (4D), as demonstrated in Chapter 5, "BIM and Construction," involves synchronizing the model components with the schedule to visualize the sequence of construction. But does this really bring value to the field staff? Will it accelerate the construction process? Does it mitigate risk? How is it used? When is it updated?

Software companies are a for-profit organization. Therefore, their products will always claim to be the most revolutionary *technology* to streamline *process*, but model-based scheduling is no different than any other BIM tool and additionally requires a *behavioral* change to be a success. If the processes and behaviors are foundationally flawed, then don't expect the technology to miraculously be the answer. Our industry has not been setting a good example in regard to meeting deadlines. Our average hit rate is 55 percent during design and 60 percent during construction (Figure 6.15). Most teachers would probably say we're behaving like unorganized and distracted teenagers.

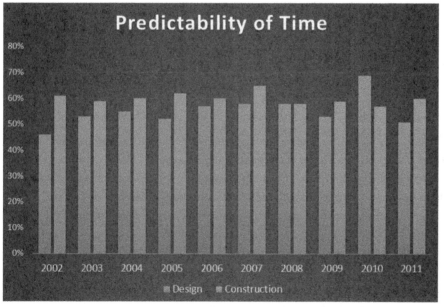

SOURCE: 2011 *UK INDUSTRY PERFORMANCE REPORT*, GLENIGAN, CONSTRUCTING EXCELLENCE, AND DEPARTMENT FOR BUSINESS INNOVATIONS AND SKILLS

Figure 6.15 Predictability of time

Chapter 4, "BIM and Preconstruction," discussed the complex issue of scheduling a cyclical process (design) with a linear method (Critical Path Method), but what is causing the linear process of construction to fail? The issue is understanding the interdependencies of the overlapped linear tasks and achieving predictability. The question then becomes, how do you achieve predictability? The answer can be found in the teachings of Frederick Winslow Taylor and his scientific studies. In order to achieve predictability, you have to behave differently. You need to have flow and consistency in your daily tasks, which is not a strength of the CPM.

For complex projects, CPM scheduling is well suited for analyzing milestones, creating logic between activities, and determining the critical path, but it doesn't necessarily give consistency in predicting the weekly or monthly look-ahead for crew activities. CPM is more reactionary, because of its high-level overview of the project; delays aren't discovered until individual tasks fail to be completed. You may have thought I was kidding about the "distracted teenager," but in his 1911 monograph *The Principles of Scientific Management,* Taylor refers to the general contractor's use of CPM schedules like an inefficient teacher giving a class an "indefinite lesson to learn." Taylor further explains that each day teachers should lay out clear-cut tasks for students to achieve, and that only by these means can systematic progress be made. This is why pull-planning is becoming a popular technique to use in tandem with CPM.

> *All of us are grown-up children, and it is equally true that the average workman will work with the greatest satisfaction, both to himself and to his employer, when he is given each day a definite task which he is to perform in a given time, and which constitutes a proper day's work for a good workman. This furnishes the workman with a clear-cut standard, by which he can throughout the day measure his own progress, and the accomplishment of which affords him the greatest satisfaction.*

Frederick Winslow Taylor, *The Principles of Scientific Management,* New York: Harper and Bros., 1911

Another method that's growing in popularity, because of CPM's weaknesses, is location-based scheduling (LBS). This method focuses on individual crews and their production rates instead of general activities. It aims to remove barriers and create consistent flow of crews from one location to another. LBS uses locations, material quantities, and production rates to create an optimal flow of crews, shown as flow lines (Figure 6.16). If lines overlap, that means two crews are in the same location, which triggers an issue. This method is more proactive, because it forecasts "what could happen" based on the average crew performance as opposed to "what did happen." This method works well for horizontal construction, high-rise buildings, or multifamily housing where tasks are repetitive. But this method can have varying success on complex projects.

Model-based scheduling programs typically fall in one of these categories (CPM or LBS), which means the behaviors required to use them and their capabilities are predetermined by the very nature of the methodology. A software program like Navisworks or Synchro utilize the CPM, which means they are best suited for high-level activities on complex projects: site logistics, steel erection, exterior skin coordination, and the sequencing of large MEP systems. This allows for large sequencing and safety issues to be discovered early, which can mitigate risk and potentially increase the schedule. It also means that it provides value when models aren't fully developed, because its value is not dependent on a high level of development (LOD).

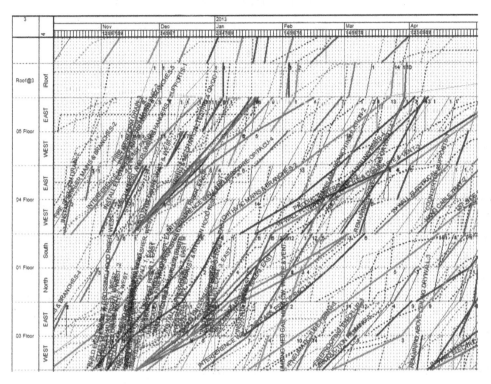

Figure 6.16 Flow-line schedule

Note: Using a CPM-based program will provide diminishing returns when leveraged for daily activities.

Vico, which is LBS driven, requires a high LOD because it has to extract accurate quantities out of the model. That means its value is more construction based after the design is complete. It improves the schedule by identifying inefficiencies and opportunities for incremental improvements. However, Vico demands more maintenance because you must make daily entries for manpower and production rates, which may require a dedicated scheduler for the project. If that information isn't maintained, the flow line is useless. When the information *is* maintained, it provides an excellent resource for the project team to understand inefficiencies in crew flow, production rates, and manpower.

Note: Using an LBS-based program will provide diminishing returns when leveraged on complex projects.

Developing BIM Intuition

Do you pay more attention to information that comes in through your five senses (Sensing), or do you pay more attention to the patterns and possibilities that you see in the information you receive (Intuition)?

Source: Adapted from *Looking at Type: The Fundamentals* by Charles R. Martin (Center for Applications of Psychological Type 1997); www.myersbriggs .org/my-mbti-personality-type/mbti-basics/sensing-or-intuition.htm

A skill that both office and field personnel must develop is the ability to see patterns and possibilities in the *information* of BIM. There's not a single tool that is the answer. All of them have strengths and weaknesses. Successful use of BIM lies in finding the patterns between the strengths to create efficiencies in achieving your goals. Developing this intuition starts with understanding where information originates and then asking, "What else can I do with it?" This is why you should develop basic skills at the element creation level before moving into analysis of any kind (coordination, scheduling, estimating, facilities management, or sustainability). If you jump to analysis prior to having foundational understanding, you may miss an opportunity to create efficiencies or not be effective at all. The purpose of this next exercise is to develop your intuition by looking beyond the tool in front of you. It's designed to generate curiosity in digging deeper into the possibilities and asking, "What else can I do?"

Note: To complete this exercise, you'll need the following software:

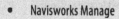

- Autodesk Revit
- Assemble Systems
- Navisworks Manage
- BIM 360 Glue account
- BIM 360 Field account
- iPad with the BIM 360 Field application

Starting with a Door

We begin by looking at the creation of a single door within Revit. This process is pretty straightforward:

1. Draw a wall.
2. Select a door family.
3. Hover over the wall.
4. Determine the swing direction.
5. Left-click to place the door.

These are sensible steps, but they can be intuitive steps. Let's dig a little deeper. When an architect creates a door, it automatically generates a useful piece of information that is hidden from the typical properties. The door associates information with the room it's placed in, but you wouldn't be able to notice it until you start to analyze the model in other platforms. In order for the door to associate with the room, the architect must place the door so that it swings out from the room (Figure 6.17). Once the door is placed, the architect can adjust the swing however they desire, but it must initially be placed swinging out. You'll see the importance of this later in the exercise.

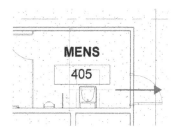

Figure 6.17 Placing the door so that it swings out

After the door is placed, architects usually tag the door with a name that corresponds to the room they're servicing (Figure 6.18).

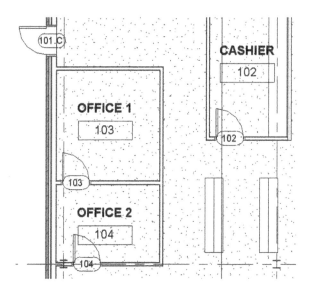

Figure 6.18 Rooms and door tags

Adding a tag fills in a field under Identity Data in the Properties of that door instance. You'll notice it under the Mark parameter (Figure 6.19).

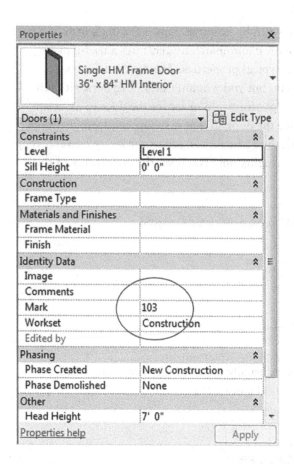

Figure 6.19 Properties window in Revit

Tagging components within the model allows the architect to quickly generate schedules for those components, which saves them time compared to the manual entry of unintelligent objects. Who else could benefit from this information? What else could you do with it? One individual who would be interested in automating component information is the project estimator.

Assemble Systems: Beyond the Basics

As you learned in Chapter 4, one person who benefits from the architect's use of BIM on the contractor side is the estimator assigned to the project. Estimators can automatically extract the quantities of the doors or other components using Assemble Systems.

1. Download the model Door Tracking.rvt.
2. Upload the model to Assemble Systems using the steps from Chapter 4.
3. Use the Filter Instances By options to isolate all the doors on the project (Figure 6.20).

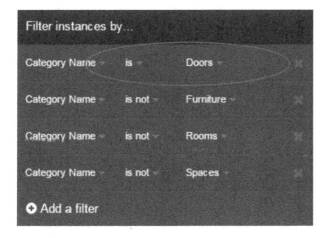

Figure 6.20 Filtering categories in Assemble

This filter can be used to isolate any element in the model (Doors, Walls, MEP Equipment, etc.). The sensible thing about this tool is that it allows estimators to quickly extract and compare quantities out of the model, but there is a very intuitive tool within Assemble that's not typically advertised. In Chapter 5, you learned how to create search sets within Navisworks Manage to tie groups of elements to TimeLiner, but separating the elements with the Find Items tool can be time consuming for both novice and experienced staff. In the bottom-left corner of Assemble, you'll find an Export Navisworks Search Sets option (Figure 6.21). This feature makes it easy for anyone to create search sets. It will automatically generate search sets based on whatever is isolated in the 3D view.

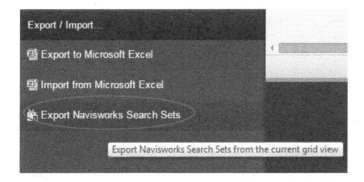

Figure 6.21 Exporting Navisworks Search Sets out of Assemble Systems

4. Make sure the doors are still isolated, and click Export Navisworks Search Sets.

5. Deselect Include Instances and save the file as **Door Tracking Search Set.xml** (Figure 6.22).

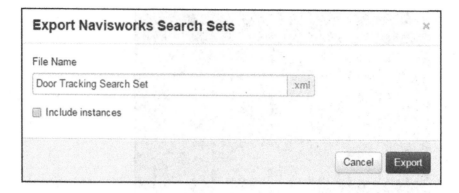

Export Navisworks Search Sets   ✕

File Name

| Door Tracking Search Set | .xml |

☐ Include instances

Cancel   **Export**

Figure 6.22 Saving the search set

Importing Search Sets into Navisworks

Before continuing with the exercise, download and install the Autodesk Navisworks NWC File export utility for Revit:

www.autodesk.com/products/navisworks/autodesk-navisworks-nwc-export-utility

1. Navigate back to a 3D view in Door Tracking.rvt.

2. In the top ribbon choose Add-Ins ➪ External Tools ➪ Navisworks 2015 to export an NWC file (Figure 6.23).

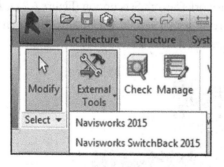

Figure 6.23 NWC Export

3. Save the NWC file.

4. Open Navisworks Manage.

5. Use the Append button found in the top-left corner of the ribbon to open the NWC file.

6. On the Home tab, choose Sets ➪ Manage Sets to open the Sets window (Figure 6.24).

7. Right-click in the Sets window and select Import Search Sets.

8. Import the Door Tracking Search Set.xml file.

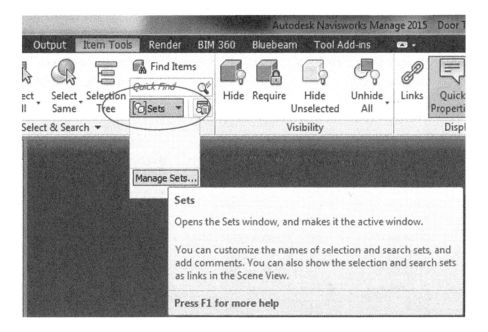

Figure 6.24   Manage Sets

You'll notice that the search sets are sorted by type and not by instance. This was accomplished by deselecting the Include Instances check box in Assemble before exporting the search sets (Figure 6.25).

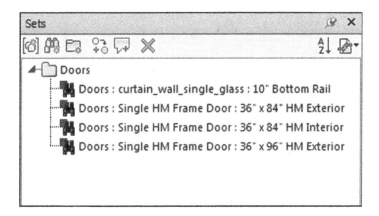

Figure 6.25   The Sets window in Navisworks Manage

9.   On the Home tab, make sure the Properties window is visible.

10. Right-click on any door in the model and select Set Selection Resolution To First Object.

11. Select a door.

12. In the Properties window, select the Element tab.

Note all the information about the door that was extracted from Revit, including Mark and an interesting field called From Room. This was the field populated by the swing of the door and can be useful for the field staff tracking where doors are located on a project (Figure 6.26).

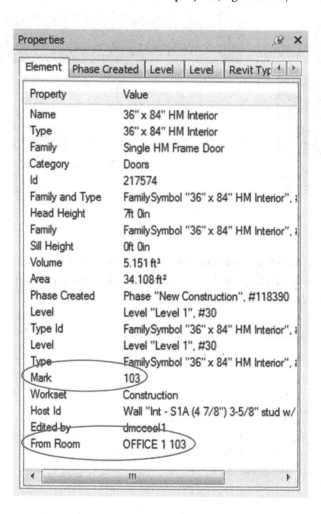

Figure 6.26 The Element tab of the Properties window

Search sets can be used in Navisworks for a number of sensible tasks, including clash detection, scheduling, animations, and color coding the model with the Appearance Profiler tool. Additionally, they have a more intuitive function with their ability to let you automate equipment mapping in BIM 360 Field to mobile devices for tracking materials, punch lists, and commissioning.

Mapping Equipment to BIM 360 Field

There are currently two ways to map equipment into BIM 360 Field, but both require BIM 360 Glue as a conduit. One method is creating equipment sets manually in Glue. You right-click on any object and select Create Equipment Set. This acts like a type search set and will group all the instances of that type as a piece of equipment. For example, you could click on a mechanical pump in 360 Glue and create an equipment set, and all the similar pumps would be grouped into this one set. A more intuitive method, which allows more control, is to upload models directly from Navisworks Manage with equipment isolated into search or selection sets.

Note: Before continuing with the exercise, be sure to download and install the Autodesk BIM 360 add-in app for Navisworks Manage and Navisworks Simulate from https://b4.autodesk.com/addins/addins.html.

1. After installing the BIM 360 add-in, you'll see a new tab in the Navisworks Manage ribbon called BIM 360.

2. Select this tab and select Glue to upload the Navisworks model to Glue.

3. Open the desktop 360 Glue application and open the uploaded model.

4. In the toolbar, select Models ⇨ More Actions ⇨ Share With Field (Figure 6.27).

5. Open the BIM 360 Field desktop application.

6. Click the drop-down menu in the top-right corner and select Setup (Figure 6.28).

   This is the back-of-house setup for the administrators of a project in 360 Field. It allows the administrator to customize what the project teams have access to and can see on mobile devices or on the desktop application.

7. Locate Equipment in the toolbar ribbon.

8. On the Models tab, select Add Model From BIM 360 Glue and upload your Glue model.

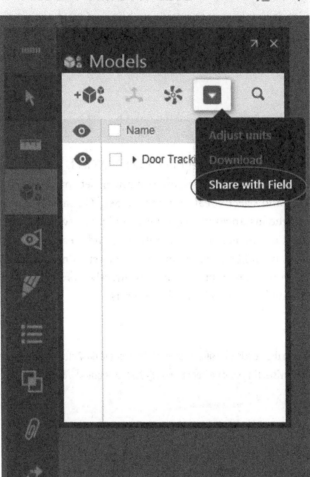

Figure 6.27 Share With Field

9. On the Types tab, create a new category called **Architectural** and create a new type called **Doors**.

10. On the Statuses tab, delete everything (this step may not be allowed with the trial version) and then add four statuses: No Status, Delivered, Installed, and Damaged.

11. On the Standard Properties tab, make sure the check boxes are selected for Barcode, Submittal, Warranty End Date, and Warranty Start Date.

12. On the Custom Properties tab, make sure Manufacturer is listed. If not, add it as a custom property.

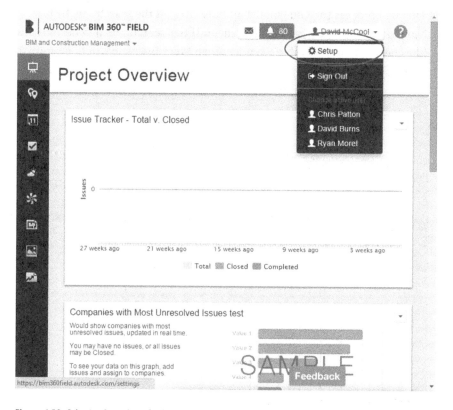

Figure 6.28 Selecting Setup from the drop-down menu

Now you're ready to map the equipment.

13. Go back to the Models tab and select Manage Equipment Mapping (Figure 6.29).

Figure 6.29 Selecting Manage Equipment Mapping

This will process the Glue model and discover all the search sets from Navisworks. Notice how it groups them. If our search sets had been by Instance (the check box in Assemble's export), then every instance would be shown in the Manage Equipment Mapping screen. Because our sets were by Type, they group together quite nicely (Figure 6.30).

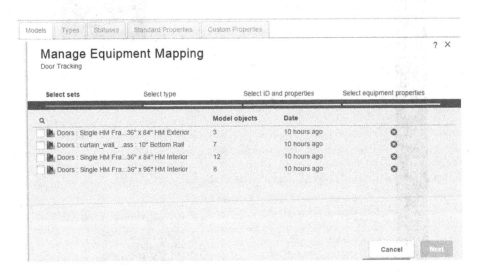

Figure 6.30 Our sets group together nicely.

14. Select all the search sets and then click Next.

15. Use the drop down to locate the Doors type that you created in step 9 to classify these sets as doors and then click Next.

16. Click Advanced in the Choose Mapping Mode window and then click Next.

17. Select the Show All Categories check box and search for Mark; then click Next.

    The Mark category is the same "Mark" that the architect put on the door in Revit. Remember, it's not only the door tag—it's the room number. In the next window, we're able to associate the Standard Properties we set up with the parameters from Revit.

18. Search for From Room and associate it with Description (Figure 6.31).

19. Search for Mark and associate it with Barcode.

20. Click Next.

21. Select the following properties: Status, Submittal, Manufacturer, Warranty Start Date, and Warranty End Date.

    Note: You'll need to click Select Property each time.

22. Click Next Set and repeat steps 15–21.

23. Once all four sets are complete, click Save Mapping.

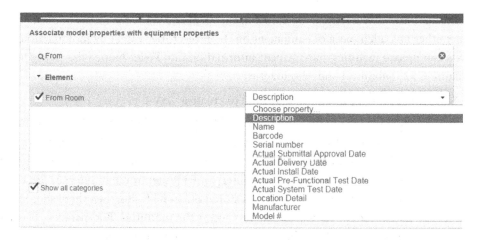

Figure 6.31 Associating model properties with equipment

24. Navigate back to the BIM 360 Field dashboard by clicking the B in the top-left corner.

25. Click the Equipment icon in the toolbar on the left-hand side.

26. Choose More Actions ➪ Customize View to match the fields in Figure 6.32.

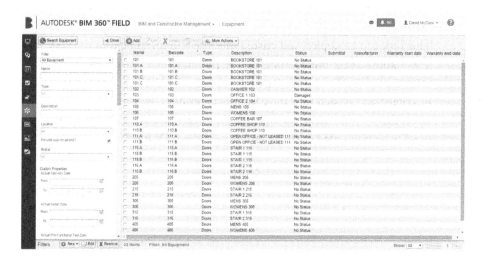

Figure 6.32 Equipment database in BIM 360 Field

Information Loading and QR Coding

The Equipment window is a direct link to push/pull information from the model and can be shared with mobile field applications to create efficiencies for the field staff and facilities management. This database allows you to upload facilities management information, easily navigate the model in the field, and track the status of doors or any other uploaded piece of equipment on the project. Let's start by looking at how you can upload facilities management information into Field. Notice the empty fields in Figure 6.32 (Submittal and Manufacturer).

1. Select More Actions ⇨ Export All to create a CSV file.

2. Expand the CSV cells to make the Submittal and Manufacturer cells visible.

   Can you think of any submittal information you could put in this field? If you created a Digital Plan room, then it's probably saved in the 02 Submittal folder, which might be hosted on a computer, an FTP site, or on an internal project management solution. If the file is stored on an FTP site, then there should be an address that can be generated to access the submittal. For instance, Box.com will generate a code to share files by access rights (Figure 6.33).

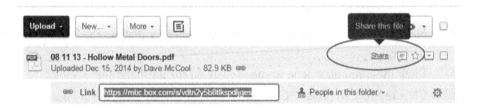

Figure 6.33 Generating a file link in Box.com

This link can be copied and pasted into the Submittal field of the CSV file. The manufacturer information can be populated by navigating to the door information on the manufacturer's website. The URL of the manufacturer can be copied and pasted into the Manufacturer field (Figure 6.34). Depending on when the project reaches completion, the warranty start and end dates can also quickly be populated in the open fields of the CSV file (Figure 6.34).

G	H	I	J
Submittal	Manufacturer	Warranty start date	Warranty end date
https://mbc.box.com/	http://www.cecodoor.com/	5/1/2015	5/1/2025
https://mbc.box.com/	http://www.cecodoor.com/	5/1/2015	5/1/2025
https://mbc.box.com/	http://www.cecodoor.com/	5/1/2015	5/1/2025
https://mbc.box.com/	http://www.cecodoor.com/	5/1/2015	5/1/2025
https://mbc.box.com/	http://www.cecodoor.com/	5/1/2015	5/1/2025

Figure 6.34 Fields in the Equipment CSV file

3. Save the CSV file.

4. Select More Actions ⇨ Import to import the added information into the model.

   So far, most of the steps in BIM 360 Field have been on the sensible side; create a database, populate fields, and so forth. But what else can you do with this information? There's a reason why I wanted you to recognize the Mark property and associate it with Barcode. Autodesk BIM 360 Field's mobile application has a barcode scanner that allows you to instantly pull up details, checklists, issues, attachments, activities, and tasks associated with a piece of equipment in the field (Figure 6.35).

Figure 6.35 BIM 360 Field's mobile application

One issue that project teams have with leveraging BIM in the field is navigating the model. Touch screens have made it a lot easier to navigate models than using a stylus, but it still can be difficult to locate yourself in a certain location to validate installation, perform quality control, check equipment status, or manage punch lists. In the top-right corner of Figure 6.35, you'll notice a link that says, "Show in Model." By selecting this link, you're automatically taken to that exact piece of equipment in the model, which in this case is Door 104 in Office 2. The Mark that the architect placed in Revit now becomes a useful solution to tracking doors in the field and navigating the model.

5. Highlight all the door tags under the Barcode field in the CSV file.

6. Copy and paste the values into a batch QR or Barcode creator like QRExplore .com to generate codes.

Be sure to add the Mark on the barcode in order to associate it with the right door (Figure 6.36).

Figure 6.36 QR code associated with the right door

101.A

These QR codes or barcodes can be placed on doors by the manufacturer prior to delivery, in the field once they arrive, or after installation during the quality control process. They provide a valuable tool throughout construction for quickly validating information in the field (Figure 6.37). See Table 6.1 for a comparison between barcodes, QR codes, and RFID tags. Currently, BIM 360 Field is not compatible with RFID scanning, but that may change in the future.

IMAGE COURTESY OF McCARTHY BUILDING COMPANIES, INC.

Figure 6.37 Superintendent using BIM Anywhere to scan QR codes for quality control in the field

Attribute	Barcode	QR code	RFID
Line of site	Required	Required	Not required (in most cases)
Read range	Several inches to feet	Several inches to feet	Passive RFID; up to 30 feet Active RFID; up to 100 feet
Identification	Most identify only type of item (not uniquely)	Can identify each item uniquely (limited up to certain value)	Can uniquely identify each item
Read/write	Only read	Only read	Read write
Technology used	Optical (laser)	Optical (laser)	RF (radio frequency)
Automation	Most barcode scanners need humans to operate	QR scanners need humans to operate	Fixed scanners don't need humans to operate
Updating	Cannot be updated	Cannot be updated	New information can be written on old tag
Tracking	Manual tracking required	Manual tracking required	No need to track
Information capacity	Very less	Less	More than QR and barcode
Ruggedness	No	No	Yes
Reliability	Wrinkled and smeared tags won't work	Wrinkled tags may work 30% data recoverable	Nearly flawless read rate
Data capacity	Less than 20 characters with linear	Up to 7,089 characters	100–1,000 characters
Orientation dependent	Yes	No	No
Marginal cost	$0.01	$0.05	$0.05-$1.00

Source: "Comparative Study of Barcode, QR-Code and RFID System," by Trupti Lotlikar, Rohan Kankapurkar, Anand Parekar, and Akshay Mohite, *International Journal of Computer Technology & Applications*, 4 (5): 817–821. Available at www.ijcta.com/documents/volumes/vol4issue5/ijcta2013040515.pdf.

Using 360 Field to Status Material

When you mapped the doors to BIM 360 Field, you created four statuses (No Status, Delivered, Installed, and Damaged). These status options allow the field team to track the doors from delivery to closeout. For example, on some projects subcontractors are allowed to use scissor lifts to install their overhead work. On occasion these scissor lift operators will damage the door frames while trying to navigate the equipment into the different rooms. A field staff person could identify this on a field walk and immediately change the status from Installed to Damaged to ensure the door gets fixed prior to closeout.

1. Download the Autodesk BIM 360 Field application to your mobile device.

2. Sign in using your Autodesk account and navigate to the project you created in BIM 360 Glue.

3. Use the Sync button in the bottom-right corner to sync your mobile device with BIM 360 Field.

4. Use the Barcode Scanner feature to scan the QR code for door 101.A (Figure 6.38).

Figure 6.38 Barcode Scanner

5. Scroll to the bottom of the page and note the facilities management details that transferred from the CSV file (Figure 6.39).

Figure 6.39 Facilities management details

6. Scroll back up to the Status field and change the status to Delivered.

7. Repeat step 6 for all doors to change the status to No Status, Delivered, Installed, or Damaged.

8. When finished, repeat step 3 to sync the mobile device with BIM 360 Field.

Visualizing Equipment Status in the Model

Open the desktop BIM 360 Field application and navigate to the equipment section. A database of missing information (Figure 6.32) has now become a robust database of material tracking and facilities management information that can be used during construction and for the lifetime of the building (Figure 6.40).

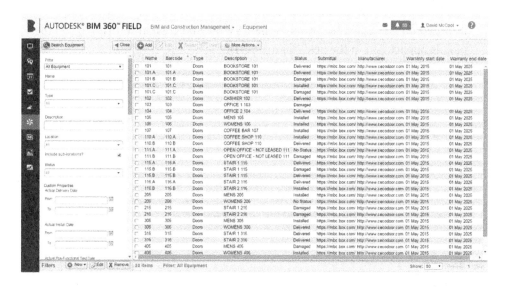

Figure 6.40 Equipment database in BIM 360 Field

The last intuitive lesson I'll discuss is how to round-trip the information database back to where it originated: Navisworks Manage. Earlier in this exercise you saw how to use search sets in Navisworks Manage. Search sets can be used in conjunction with the Appearance Profiler tool to map object information to colors. This comes in handy during constructability review. Think back to the lessons in Chapter 4 about details and "seeing beyond the clash." There were two specific areas critical to the design that require more than a click of the button: Fire Life Safety and Water Infiltration.

Architects put information in the model for the rating of walls to annotate and distinguish the difference between them in the plans (non-rated, fire rated, acoustically rated, and so forth). When these models transfer to Navisworks, the walls are generally all the same color. The Appearance Profiler allows you to map colors to these ratings and then instantly color the model with those parameters. For example, your non-rated walls could automatically be colored white with transparency, fire rated (red with transparency), and acoustically rated (green with transparency). By visualizing the walls, project teams can quickly identify Fire Life Safety issues and address them during coordination. Can you think of how this might apply to the statuses you just created?

1. Open Navisworks Manage.

2. On the BIM 360 tab, click Open.

3. Log in, select your project, and then click Next.

4. Use the drop-down menu to select Models instead of Merged Models.

5. Select your project and then click Open.

6. Navigate and select any door on the project .

   You'll notice a slight change to the properties of the door. In the properties window there is now a tab called BIM 360. On this tab you'll find all the information you uploaded through the CSV file and created statuses for on the mobile application in the field (Figure 6.41).

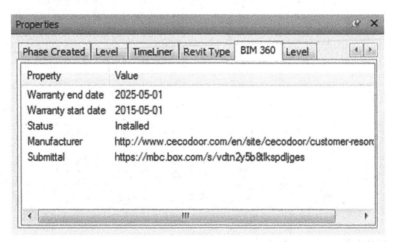

Figure 6.41 BIM 360 Properties in Navisworks Manage

This information, like the wall ratings, can be mapped to colors to visually show an overview of material and/or equipment statuses.

 Note: Use the exercise on search sets from Chapter 5 to isolate the various statuses or download and import the file Door Status.xml.

7. On the Viewpoint tab, make sure Render Style ⇨ Mode is set to Shaded.

8. Open the Appearance Profiler tool from the Home tab.

9. Select the By Set tab.

10. Click on each set, assign a color, and then select Add (Figure 6.42).

11. Click Run.

Appearance Profiler ✕

Selector

By Property | **By Set**

▷ 📁 Doors
▷ 📁 Doors (Door Status.xml)
🗂 Damaged
🗂 Delivered
🗂 Installed
🗂 No Status

[Refresh]

[Test Selection]

Appearance
Color   Transparency

[▮▮]   ○─────────   [0] %

Selector	⬤	🗂
Damaged	▮▮▮	0
Delivered	▯▯▯	0
Installed	▮▮▮	0
No Status	▮▮▮	0

[Add] [Update]   [Delete] [Delete All]

[Load...] [Save...]   [Run]

Figure 6.42 Appearance Profiler settings

All the doors will instantly change to the colors you assigned to them. In the Selection Tree window you can highlight all the search sets, right-click, and isolate the doors by choosing the Hide Unselected option. This will give the project team a quick status overview of all the doors on the entire project by color (Figure 6.43).

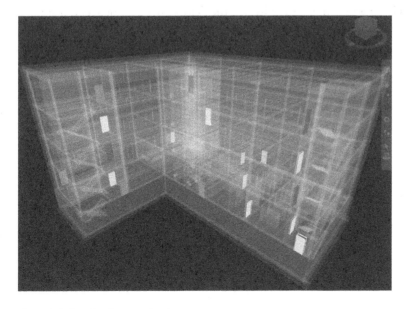

Figure 6.43 Overall project status by color

Endless Possibilities

This exercise demonstrates the power of information in BIM. The model is just a database that can be manipulated and leveraged in multiple ways throughout design, construction, and turnover. This chapter only scratched the surface of the value models can bring to the field staff, but hopefully it got you thinking about ways you can leverage the information on your project. If you understand the patterns of how software translates information, then it opens up endless possibilities. Let's recap how the information translated throughout this exercise (Figure 6.44).

- Revit created the door tag and location.

- Assemble was used to isolate the doors and quickly generate search sets, which is more efficient than manually doing them in Navisworks Manage.

- The search sets were imported into Navisworks Manage to automate equipment mapping into BIM 360 Field, which is more efficient than creating equipment sets one by one in BIM 360 Glue.

- Glue was strictly used as a conduit from Navisworks to Field.

- BIM 360 Field created a database to easily import facilities management information into the model using the CSV file.

- Field leveraged the door tag information to generate custom barcodes by room number to stick on doors, so the entire project team could easily navigate and use the model in the field for punch lists, quality control, and installation verification.

- The Field mobile application helped the field staff verify delivery, installation, and quality of the doors in the field.

- Navisworks Manage was used to visually summarize the status of the entire project.

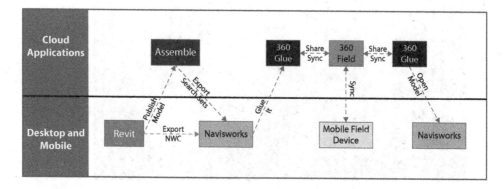

Figure 6.44 Flow of information example

Small Wins to Big Change

For a company to become BIM-enabled, they must treat every construction project as an opportunity to develop innovative ways to leverage BIM. The thing that's so fun about our industry is that no project is exactly the same. Each one presents unique challenges to overcome and BIM may provide the key to overcoming them. Equipped with BIM tools, field staff must be encouraged to fail often and fail fast in facing these challenges. That's the only way to find the small wins and start chipping away at the barriers of adoption. Once the successes are discovered, they need to be celebrated and communicated. This will do four important things:

Boost Morale Everyone likes to know they did a good job and to feel appreciated. When someone discovers a new way to use BIM, it should be viewed as a chance to promote your project and encourage your team. Find ways of communicating the small wins to the company, whether it's an internal e-mail, website, YouTube, or other means. This reinforces to the field staff that what they're doing is important to the company initiative and that they're making a difference. Additionally, it will create a desire to discover more ways to leverage BIM tools.

Bring Consistency Construction projects are similar to a small business. They have their own address, phone number, staff, client, mission, and budget. Sometimes projects become isolated from the rest of the company because of this fact. Communicating the successes can help bring consistency to the company and prevent other projects from struggling through the same pitfalls. Even though projects are unique, there are a lot of common issues that all of them face.

Break Down Barriers There will be skeptics and those clinging to the traditional past even after the successes are revealed. However, the more small wins are discovered and communicated, the less they will have to hold on to.

Build Momentum The small wins will build momentum as they chip away at the old process. In *The Principles of Scientific Management*, Fredrick Taylor states that the threshold for companywide adoption is 33%. Once you reach that percentage of BIM use, the remaining laggards and skeptics will become eager to jump on-board. They'll see the benefits gained by the new process and won't be able to ignore it any longer.

Summary

This chapter covered BIM and the battle for acceptance, and showed you how to train your field staff, create efficiencies in document control, and create organizational change. You also learned the value of 4D and how to think intuitively about the "I" of BIM. All of these topics are specific to BIM processes but are not by any means the only tools that can be used on a jobsite. This chapter covered the basic concepts of BIM during construction administration, and it shows where there are current solutions and shortcomings alike.

Use these exercises as a foundation to create your own innovative ideas. Don't view the methods that were used as the answer; view them only as experiments. The goal is to expose some of the exciting ways BIM can be used in the field and to start breaking down the barriers to adoption. There are too many efficiencies to be gained to sit idly on the sidelines. Field personnel have to start challenging the status quo and learning to learn.

The digital plan room is complete, doors have been installed, equipment information has been uploaded, and the owner is getting ready to take possession of the space. So now what? The value of BIM technology lies in its ability to house important life-cycle data that offers information to all stakeholders. The next chapter will demonstrate more ways to use mobile devices for closeout and to set up the model for facilities management.

BIM and Close Out

This chapter will cover how to deliver BIM content for the facility management phase of a structure. This section will also describe why the delivery of useful and meaningful information to construction consumers is becoming increasingly important.

True Costs of Facility Operations

Designers and contractors need to know that, when it comes to total facility costs, their combined efforts only account for 15 percent of a project's overall cost in 2011. (http://buildipedia.com/aec-pros/facilities-ops-maintenance/life-cycle-view-total-cost-of-ownership-drives-behavior.) *The Long Term Costs of Owning and Using Buildings* study (Evans, Raymond; Haryott, Richard; Haste, Norman; Jones, Alan, Buildapedia, 2004) found that this number could be even less—on average, only 3 percent of a building's total cost is spent in the design and construction phase of a project. The other 85 percent is required to operate and 12 percent to maintain a structure (Figure 7.1). These maintenance costs are represented in the following ways:

- Utility costs
- Capital costs
- Insurance costs
- Maintenance and cleaning
- Equipment repair and upkeep
- Document and asset management
- Taxes
- Ongoing operational expenses

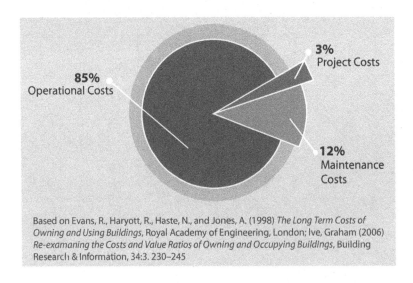

Based on Evans, R., Haryott, R., Haste, N., and Jones, A. (1998) *The Long Term Costs of Owning and Using Buildings*, Royal Academy of Engineering, London; Ive, Graham (2006) *Re-examining the Costs and Value Ratios of Owning and Occupying Buildings*, Building Research & Information, 34:3. 230–245

Figure 7.1 Life-cycle costs of a building

Other factors such as the rising cost of resources like natural gas, electricity, material and equipment prices, and labor create challenges for owners. Building occupants are speaking with their capital and demanding sustainability be integrated into the buildings they lease or purchase. Structures built for manufacturing must now be designed to maximize flexibility for production configurations while minimizing operational costs and energy use, as well as pollution reduction. The pressure to deliver high-quality, flexible, and sustainable structures into the built environment will be under increasing scrutiny because of the nature of how they are an integral part of people's lives. Many architectural, engineering, construction, and operations (AECO) professionals don't understand the potential negative impacts buildings can have. So how can you change these statistics?

- Buildings consume 40 percent of the total energy in the United States (source: U.S. Green Building Council, http://www.usgbc.org/articles/green-building-facts, 2015).
- Energy use accounts for 30 to 40 percent of the operational costs of a building (source: Energy Star, 2015, www.energystar.gov/buildings/about-us/how-can-we-help-you/build-energy-program/business-case/10-reasons-pursue-energy-star).
- Inaccurate or unorganized building data accounts for 60 percent reduction in staff productivity (source: NMSU Research Study).
- On-time maintenance can create an 8 to 15 percent operational savings per annum (source: BIM for Facility Managers, 2012, www2.deloitte.com/content/dam/Deloitte/us/Documents/consumer-business/us-avitran-thl-smartermro-072612.pdf).

You must challenge the industry norms that say a structure's operational costs can't be planned for and that binders and binders of unusable facility data are acceptable. You must promote the use of BIM for facility operations to create significant life-cycle savings. Unfortunately, many contractors fail to understand the costs involved with the operation of a facility or that many of the issues caused during operation start with the quality of as-built information delivered at project completion.

Traditionally, builders have treated project closeout documentation as a necessary evil. It becomes a "check the box" component of information that needs to be handed over to a customer to complete a contract. However, facility managers rely heavily on as-built information to understand what was built in a facility to understand exactly what was installed in order to perform their job function best. For example, I have heard numerous horror stories from facility managers who go to do maintenance, such as replace an air filter. Based on the as-built information, they remove ceiling tiles or close down a portion of a facility only to find that the equipment is in a different location or doesn't exist at all. Imagine being the captain of a ship and trying to steer based on map information that is "kind of close" or "semi-accurate." It wouldn't take long to run aground or hit a reef that was missed on the map because it wasn't treated with a high level of care and detail. I'll bet if the map makers knew how their map was going to be used, they would be sure to deliver a better quality product. Stephen R. Pettee explains the dynamic issue of quality as-built data well, in more specifics, in his whitepaper for the Construction Management Association of America (CMAA), "As-builts—Problems & Proposed Solutions" (source: https://cmaanet.org/files/as-built.pdf).

So how can construction managers better deliver project close out information to satisfy new owner requirements? Successful information delivery starts in the project planning phase (see Chapter 2, "Project Planning"), but a number of issues contribute to low-quality project turnover data. These can include project fatigue, improper planning, a lack of clear expectations, inaccurate change documentation, and insufficient staff. The issue is one that persists as a major pain point for facility owners around the world. One of the major issues with as-built quality is the technology and formatting used, which is counterproductive to the information that is needed.

For a moment, let's empathize with the role of the facility manager. Given a choice, would you rather use static and inaccurate 2D data to do your work and update, or would you rather use an accurate 3D model to find what you need, which can be updated? As Albert Einstein said, "The significant problems we face cannot be solved by the same level of thinking we were at when we created them." (*The New Quotable Einstein* (2005), Alice Calaprice) So let's explore a better way of delivering as-built information and introduce the concept of constant and artifact deliverables. This concept is a means to deliver both the story for a structure, or *artifact* (2D information sets), while still delivering a means of updating facility information both in model geometry form and as a dataset (BIM), or constant, for more effective facility maintenance.

Supporting Organizations

Virtual Builders (www.virtualbuilders.com), Fiatech (www.fiatech.org), the International Facility Management Association (IFMA) (www.ifma.org), and the buildingSMART Alliance (www.nibs.org/?page=bsa) contain reference material and support industry initiatives, tool development, and case studies that further industry discussions for the integration of BIM into facility management.

Artifact Deliverables

So what is an artifact in the context of facility management? Many facility managers may think of an artifact in reference to the Indiana Jones movies, where the hero must make it through perilous traps and life-threatening situations to get to the document they need (Figure 7.2). Although some of the searching may be similar, artifacts as discussed here are located in far less exotic locales, but they tell a story nonetheless. This story consists of the stream of 2D information that details how a project was built. This information can include specifications, submittals, changes, RFIs, record documents, and other traditional deliverables. The constant and artifact deliverable strategy recognizes and facilitates the current use of 2D static data as a means of delivering work and understands that artifacts such as photos, videos, and other dated information (while a representative "snapshot" in time) are useful.

IMAGE COURTESY OF TUSTIN UNIFIED SCHOOL DISTRICT

Figure 7.2 Documents to manage in facility management

Artifact deliverables can be defined as static project documentation. This information is in a number of forms such as printed files, PDFs, CAD drawings, and digitized communication records such as e-mail archives or official correspondence. These forms of information represent a record or history of a project's development that resulted in the final as-constructed product. The key factor in artifacts is that they lack the flexibility to change over time and are not an effective means of updating information throughout the life of a facility. Put yourself in the shoes of a facility manager who needs to reflect a change such as a removed wall. How does he or she change the documentation they rely on?

PDF Artifacts

PDFs are great. They have in many ways unified and standardized the deliverable type for the construction industry. The Portable Document Format was originally developed by Adobe Systems as an ISO-maintained means of facilitating electronic document exchange.

Thanks to readily available PDF viewers, this format has become the gold standard for exchange in the construction industry because of its accessibility and file integrity.

Note: Additionally, PDF files can be connected to each other through the use of hyperlinking as outlined in Chapter 6, "BIM and Construction Administration."

During construction, PDFs typically go through a series of exchanges and revisions, resulting in drawings that show areas in question, changes, or other data points (Figure 7.3). Though useful in its use and accessibility, to facility managers as-built PDFs are challenging to interpret for two reasons:

- The first is that the facility manager is often not part of the construction team and doesn't have the same insight into what was built as the construction personnel on-site.
- The second is that the drawings do not necessarily contain the information that they need to perform their activities. This is because the intended use of the documents was to construct a facility, not operate it.

The purpose of record drawings (the initial drawings created for a project) is to give enough information to build a project. From these record drawings, as-builts show what was built and contain the backdrop to how the project was constructed. The challenge for facility managers is that these PDF drawings were not made for the sole purpose of operations and maintenance of a facility.

There is a significant difference between drawings made to build a structure and drawings made to maintain it. Some efforts have been made by facility managers to use CAD files to cobble together maintenance drawings for operation of their facility. I will discuss CAD files, another facility artifact, next.

Currently, the majority of facility and operation personnel rely on PDFs stored digitally on servers, printed, or saved and filed for reference. Revised or clouded PDFs often are delivered to an owner preloaded with all sorts of markups, notes, and other information. As time goes on, PDFs used to maintain a facility's history can quickly become overcrowded. Additionally, PDF referencing requires a significant amount of time and attention to get to the *right* document(s) and to understand the history and the changes about a facility to find the information that is needed to perform a task. PDFs for facility management create somewhat of a one-way street once embarked upon, because the facility manager has chosen the method of sharing and updating information that doesn't allow for native file editing to move walls, update equipment schedules, or show current conditions graphically. Instead, there are layers upon layers of clouds, comments, or notes that represent the changes that have been made over time and after significant use become essentially unusable.

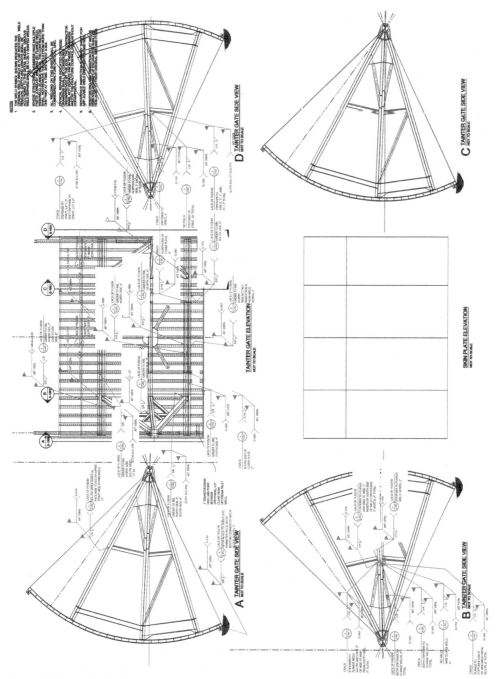

Figure 7.3 Revision plan sheet

CAD Artifacts

In the past, CAD files have been distributed to the facility manager upon project completion. Paid for by the owner as part of the contractual project requirements, these files were included for the design and construction team as a means of reflecting what was built in a new and better way. These documents then became the digital means of reference for the operations of the facility. If we hit the pause button right there, two things should stand out. One may be, "Aren't we going through that again now?" and the other may be, "So why are owners still asking for CAD files?" Good questions. At one point owners began asking for CAD files in addition to the printed files on a project because the technology had improved to a point that owners could clearly see the value. For example, CAD represented a means of updating the native files of a facility. This meant that changes and future improvements could be graphically updated by owners in CAD files that they owned and maintained.

The maintenance and use of CAD files in facility management also presented a number of challenges, among which were the addition of new processes and costs. These costs came in the form of new staff or third parties to manage CAD files, costs of the hardware and software to support their use, and disruptive processes to the way files were updated previously. In addition, these as-built CAD files weren't ready for the facility manager to use out of the gate because of their intended purpose to construct rather than operate. For example, square footage of rentable space versus gross square footage required that the rooms be traced or "polylined." *Polylines* are closed loops consisting of line and arc segments that are used to define a room and area so that information can be assigned to it. So the initial drawing conversion process was somewhat tedious as staff traced each of the rooms for use in operations and in some of the newly emerging CAD-enabled facility management or computerized maintenance management system (CMMS) tools.

If CAD files can be updated, why are they considered an artifact? Simply put, CAD files are often too numerous and disconnected to successfully enable updates to the repository of digital information effectively. Although updating a floor plan may not be too painful, a host of other related documents—including reflected ceiling plans, schedules, enlarged plans, details, and other drawings—need to be updated for each change. CAD files can easily run into the hundreds or even thousands depending on the project. CAD drawings are separate files that can be linked into a parent view for reference to manage the facility.

Next, updating CAD files is quite time-consuming. In CAD, a change in one place does not persist as it does in BIM, and updating requires each of the related CAD drawings associated with the change to be opened, altered, and saved. Another downside to CAD is the lack of intelligence within the files. For example, in BIM a wall will contain all the unique parametric characteristics that define it—such as what it is made of, its height, its area, its volume, and so on—which can help facility managers quickly drill down to the information they are seeking. By contrast, in CAD the walls are represented by lines, and the amount of intelligence associated with them is information such as colors, layers, and plot weights. Thus, CAD files must be used in tandem with specification information, operations and maintenance manuals, and the contract documents.

It's important to note the lack of flexibility within CAD information. Facility managers are often responsible for maintenance, event scheduling, setup and cleanup of the building, and tenant management, including move management and data management for a building. What often happens is that facility managers, because of the time demands of upkeep in a building, rarely maintain the digital CAD documentation that is required. As a result, CAD files used by the facility manager often become a plan representation of the facility, and the other drawings go by the wayside. As tenants move, repairs are made, facilities are upgraded, equipment is switched out, and facilities change use. The management of the data is the facility manager's alone. Typically, there is no involvement by an architect, engineer, or general contractor to document these facility changes that take place.

The last hurdle in CAD as it relates to facility operations is the accuracy of the CAD files. During a CAD process, field changes, minor alterations, and other changes are rarely documented in the CAD files. Creating true, as-built CAD data is not viewed as an important part of the architect's or engineer's scope of work. As a result, the information that the facility manager gets from the beginning of the facility's life cycle is most likely dated and inaccurate.

Constant Deliverables

The second part of the artifact deliverable is the constant deliverable. Contrary to their counterparts, constant deliverables are fluid, can be easily updated, and intuitively link to other facility information. As a means of updating information, both graphically and linking to static artifacts, BIM is an invaluable tool as a constant deliverable.

Unlike its predecessor, CAD, BIM represents a way of consistently updating facility information without having to find and revise multiple CAD files. Additionally, BIM can more easily be custom configured to the uses of a facility manager. For example, model viewing filters can be applied to see only maintainable objects such as mechanical, plumbing, or communication systems, similar to the door status colors shown in Chapter 6, "BIM and Construction Administration," Figure 6.43.

Another, more dynamic factor is to think of the delivered facility as a database of intelligent components that can be linked to or used to more efficiently populate maintenance records. Combined with an artifact management strategy, BIM creates a dynamic means of capturing and updating life-cycle facility information for a project quickly (Figure 7.4). In BIM, a facility manager can update wall information, ceiling or lighting layouts, or floor finishes, and such changes persist throughout each of the model views.

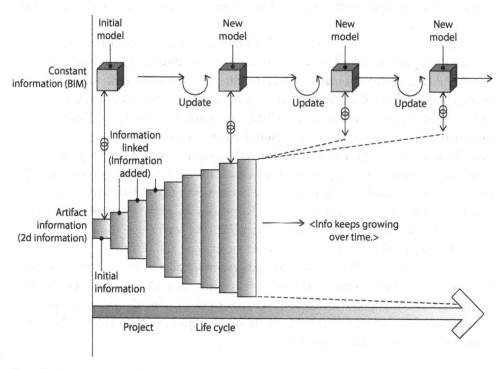

Figure 7.4 Constant and artifact information management strategy

It's All There

I like to use the comparison of an intelligent, parametric model with the delivery of boxed furniture that needs to be assembled. Typically, in the instruction manual there is an itemized breakdown of each of the components you should have in order to build the table or chair. This detailed list makes it easy to quickly determine whether or not you can complete the work.

Similarly with BIM, the discussion of model detail or "modeled to" strategies that leave out components may miss the boat in capturing the information needed for operations, making the job of using BIM for facility management exponentially more difficult because you don't know if it's all there. For this reason, I recommend that facility managers either work to get accurate record BIM delivered at project closeout or take responsibility to refine their models once received for their own uses.

Taking a Hybrid Approach

The case for artifact and constant is simple, and the reality is that a hybrid approach to managing facility data is the most effective. Although I won't dive into the best practices for determining what information should be captured in the model and what should be an artifact, know that representative model geometries and dimensional accuracy are a must for the model to be effective in facility operations. In this approach, the model continues being the single virtual representation of a structure, with the modification of its new purpose for operations informing what components should link to.

Owners and BIM

Many owners acknowledge that they can benefit greatly from an artifact and constant deliverable strategy. In particular, owners are realizing the value of an accurate as-built model as a project deliverable—less as a 3D visualization tool and more as a means of organizing and connecting the BIM database of information to address the problem of dealing with the large amounts of static data that grow over a project's life cycle. Traditionally, dealing with facility information has been a daunting task. For example, submittals provide specific information about almost every product in a facility and whether it is for door hardware, HVAC systems, or paint colors; this required deliverable results in an extensive set of information that is often overwhelming.

The larger and more complex the project, the more submittals and thus the more information owners have to deal with. Specifications are another example of large amounts of end-of-project data. Specifications describe the products used, warranty requirements, and installation instructions. Huge amounts of unlinked information going through many hands can result in unorganized or, even worse, lost data. Some of the biggest costs associated with the current system of information handover is lost time in trying to locate information for maintenance activities. The time it takes facility managers to locate needed information about a facility is proportional to their ability to meet the facility's other needs. Each minute of time searching for information sets the facility manager that much further behind in performing maintenance activities for the facility. In some cases, warranties are voided on HVAC, roofing, and flooring systems because of improper maintenance of records or file access. This can result in the unnecessary replacement of major systems and shutting down areas or closing a facility for maintenance, which isn't a feasible solution in some structures such as certain hospitals and high-security government buildings.

Although information may be delivered in an organized format, the more often it's used, the faster it becomes disorganized. Owners stand to save considerable dollars in a BIM process that results in a record BIM deliverable for the following reasons:

- Increase in staff efficiency to get to information (time)
- Maintenance of equipment to warranty standards (risk profile and expense)

- Proper documentation of commissioning issues (life safety, fire stopping, accessibility)
- Limited amount of wasteful printing (costs)
- Ability to back up critical digital and facility data that could be lost (risk profile)
- Information embedded or linked to the model to avoid unnecessary waste
- Less chance for facility downtime as a result of improper maintenance
- More efficient repair response
- Improved client/occupant satisfaction

As-built BIM files are often referred to as a *record BIM* file. A record BIM is the single, current, composite reference model for a structure that references information about the project. The process of completing a record BIM involves planning and additional model checks throughout design and construction.

The difference between a record BIM file and a record drawing is that the record BIM may change over time as facility information is updated. This equates to significant savings in information finding and analysis, as well as providing design teams with accurate as-built information for renovations or additions to the facility. So, the questions now become, why wouldn't an owner want a record BIM at project closeout if it better equipped the operations team, and who pays for it?

Owner Options

So, what are the options an owner has for receiving a record BIM? Typically, there are two means of requesting a record BIM. In both, the terms and expected level of detail should be discussed as early as possible—ideally in the BIM execution planning phase to eliminate issues as the project nears completion. Additionally, any negotiation of fees or required deliverables are best agreed to before model work begins and even better still as a not-to-exceed (NTE) option to the construction contract.

The construction manager creates the record BIM. The first option is to have the construction manager create the record BIM for the project. The manager has a significant advantage in managing and updating the model, because field personnel and support staff are typically full-time personnel dedicated to constructing the project. As a result, they are available to edit the model. They have been constantly exposed to the changes and construction issues throughout the course of construction on-site and thus can develop a more usable model. The construction manager may use survey tools to capture constructed conditions and leverage existing subcontractor or fabricator models to create this model as well. The potential gap in this method is that construction managers are not facility managers and may not know what questions to ask as it relates to what's important to a facility operator. Additionally, costs may be associated with this request, and it's often priced out as an optional service that can be completed by the manager.

A record BIM is created as part of a design contract. The second way a record BIM can be created is as part of a design or virtual construction professional's contract. This arrangement typically means that the architect, designer, or third party must negotiate for an additional fee during the project planning phase or during the construction administration phase of a project. This method is more difficult since the staff responsible for creating the as-constructed model is somewhat removed from the work and may not catch all the changes or areas of work on-site.

The end result of a record BIM is to understand that virtual modeling and construction are not the old way of delivering "design intent" documents. Nor is it the contractor's responsibility to construct a design according to "means and methods." The reality is that BIM is becoming a fabrication model for contractors and an accurate as-built product. In his book *Virtual Design and Construction: New Opportunities for Leadership* (*The Architect's Handbook of Professional Practice*, Wiley 2006), James R. Bedrick identifies the tremendous opportunity for the entire team to better collaborate and leverage BIM as a source of additional scope, responsibility, and potentially, money.

There are other creative ways to get an as-built model. Because of the reduction in cost of laser scanning equipment, many owners are asking for phased laser scans to be completed of their facilities that they can then have converted into models downstream if needed. This approach equates to accurate point cloud files that can act as a repository for information over time, which can capture in-wall, above-ceiling, crawlspace, shaft, or other areas that become covered after construction is completed. Additionally, phased laser scanning can be used for quality control processes to test for deviation in installations as work is being put in place compared to what was designed (Figure 7.5).

Figure 7.5 Model and laser scan overlay

Other promising tools that can deliver 3D information more effectively is new photogrammetry technology like Autodesk 123D Catch and Autodesk ReCap. This technology creates a 3D model from a collection of photos of a particular focus area. Although this technology is the least expensive, it is also the least developed. Other trends such as drones and mobile scanning machines may make offering true record BIM files easier in the future.

Integration of a Record BIM

How much information is too much information in a record BIM file? You must begin with its intended use and develop a strategy from there. Just as design and construction models have specific intended uses, the facility model needs to be developed under an intended purpose. Often an owner or facility management team will be able to voice their needs and tell you what they will be using the information for and may not know what to ask for in BIM. In this collaborative approach, the BIM team becomes a trusted adviser to deliver the right solution.

As development progresses, the model and information can be tested to ensure that delivery will be successful. For example, if the facility manager wants to import BIM data into their CMMS software to speed population, the process should be tested early to ensure that it works. Often the effort involved in delivering a record BIM is a partnership between the model development team and the end users of a facility.

Delivery Method Matters

If you're not working on an integrated type of delivery method, then you may be limited in your ability to deliver a record BIM. For example, in a design-bid-build process, the ability to coordinate with the operations team to find out what information they want to see at the closeout of a project rarely happens because of the cost-focused approach to the project.

The unfortunate aspect of a design-bid-build project, in regard to project closeout and record BIM information, is that it will rarely be well-coordinated documentation and any additional scope has a price tag. Typically, this is because construction managers need to cover some of the bets they made to get the project. Unless the proposal is extremely well written and the facility manager has had an in-depth amount of input into the solicitation, chances are project closeout documents will be relatively the same quality as what was delivered before.

Information in a record BIM is much like a set of instructions. For instance, if you are assembling a bicycle, most instructions will tell you what a piece is, what its function is, and where it goes. A more sophisticated set of instructions will tell you what metal alloy composition the frame is made of and what the recycled content of the tires are. The needed amount of information varies, because in order to efficiently operate the bike, you need to know how the parts go together but might not necessarily need an in-depth description of every component's chemical composition. However, such details might be needed if you were constructing and operating something more complex because the set of instructions would probably be significantly more complex in level of detail and instruction.

Some owners do not require that all the information in their facility be modeled because some components may not be critical to the effective operations of a facility. For example, does a power station maintenance technician need to know the type of exterior wall construction used? Maybe or maybe not. This is a discussion for each facility owner to have with their team to determine what information needs to be accurate and detailed to do their work. Many owners make it a best practice to say if it's built, then it needs to be modeled. Ultimately, this strategy has the most potential for downstream use, but it requires a larger investment in lieu of a hybrid strategy.

The record BIM needs to contain any information that is critical to operating efficiently without spending unnecessary time and effort modeling information.

Know the LOD

When defining LOD at the beginning of the project, carefully consider what the LOD means to the team, especially facility management. LOD doesn't just define the quality of the model—it also defines what information is contained in the model. It can get confusing. I've mentioned the unrealistic blanket statement of "models will be delivered at LOD 400 (fabrication level)," but I also need to clarify issues of information associated with LOD 400, as defined by the AIA.

The AIA's definition of LOD 400 can be construed to create a lot of effort for designers and contractors while providing little to no value for owners. The definition implies that the model will contain information for fabrication, assembly, and installation. It goes on to state that the model can be relied upon for analysis, estimating, scheduling, and coordination. Those are pretty broad statements. You could interpret this to mean that every door hinge has instructions for assembly and costs associated with it. To take it to the extreme, it could mean that the screws have costs associated with them.

Know what LOD definition you're using and clarify the information required. Don't brush over the LOD matrix. Define it in the addenda, execution plan, or the comment/notes column of the LOD matrix.

Future Owner Challenges

Typically a building's operating costs and management are regarded as necessary expenses, not areas that could provide potential savings. Additionally, the high environmental costs associated with reducing pollution emissions and vacancy costs related to poor overall occupant health are hard facts that face today's facility owner. A FacilitiesNet survey found in 2008 that more than 80 percent of facility owners are budgeting for green initiatives and energy savings in their facilities (source: www.facilitiesnet.com/news/article.asp?id=10192). The opportunities to design high-performing buildings that are more efficient, built to healthier standards, and maintain the best possible operating standard are critical in today's economy. The ability to manage the triple bottom line of social responsibility, environmental stewardship, and economic prosperity is becoming paramount, and many owners are looking to BIM and technology to offer solutions to create savings—especially if you consider the impact even a small annual savings would have when multiplied by the years a building is in use.

The Whole Building Design Guide found in 2014 that the anticipated costs of operating a facility have risen approximately 13 percent over the years 2012-2014 and will rise in the future (source: www.wbdg.org). Not only are energy costs rising, but there is an increased demand within the market for these healthier and more sustainable buildings from building tenants. Although this requirement doesn't reflect any direct cost, it does translate into marketability. New sustainable construction and retrofitting has increased drastically, and 78 percent of building owners in 2011 have said that energy efficiency was a design priority in their construction and retrofit projects; see the full article, "More U.S. employers measuring cost savings from environmental effort," here:

http://www.fmlink.com/article.cgi?type=News&archive=false&title=
Survey%3A%20More%20U.S.%20employers%20measuring%20cost%20savings%20from%20
environmental%20efforts&mode=source&catid=&display=article&id=41440.

There are other costs to tack onto this, including inefficiencies in response time, asset loss, facility management staff, and the transfer of data over time that creates waste (Figure 7.6). According to David Jordani of Jordani Consulting Group, in 2007, construction is a $3 trillion industry with 50 percent waste generated through its life cycle (source: http://bit.ly/1Dg15rl). This number equates to a serious loss of profit for building owners and, as a result, is reflected in the time and quality of a construction project as well as its operation over its life cycle.

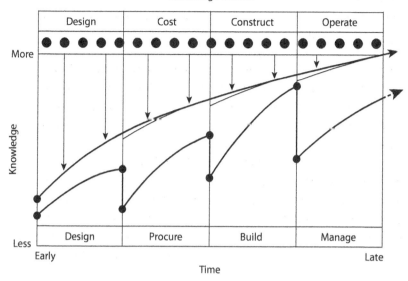

Figure 7.6 Knowledge gaps in handoff

BIM and Owner Solutions

The following is an example of potential value creation areas in the use of BIM for facility management and operations. Some of these areas are more developed than others, but there are numerous use cases for how a model can be leveraged in operations.

- Architectural programming studies
- Space functions
- Secure area analysis
- Area calculations
- Volume calculations
- Engineering performance criteria
- Specifications
- Survey information
- Change order history (potential and actual)
- Shop drawings and submittals
- Procurement files
- Progress photos/scans

- Head wall photos/scans
- Electronic diagrams
- Warranty information
- Cost information
- Purchase requests
- Work requests
- Estimates for work
- Occupant organization
- Seating arrangements
- Network diagrams
- Hazmat identification
- Operations and maintenance manuals

- Inspection reports
- Commissioning reports
- Analysis reports and simulations
- Disaster recovery plans
- Asset management and tracking

The fact is that facility management teams are finding ways to use BIM to translate into direct savings. Whether that's in time saved looking for information, response time improvement, reduction of preferred maintenance activities, better asset management, or other metrics, many owners are discovering the best means to use BIM for their purposes.

Technology recently redefined element-to-database linking through the use of RFID tags. As mentioned earlier, RFID tags are durable, small identification tags placed on assets that, using a scanner in proximity to the object, identify a piece of equipment. RFID technology is more popularly known for the *smart pass* chip that is embedded in credit cards. When a card is scanned, the user's information is directly sent to the credit card company to charge users for their purchases.

Recent developments have seen RFID tags being used in both the construction and facility management arenas. The value that an RFID tag brings to asset management that a barcode does not is that RFID tags, although small, are made out of thick plastic. Barcodes are often stickers that run the risk of becoming marred or scratched, leaving them useless. RFID tags can be placed on virtually every asset in a building and can be scanned to pull up all sorts of information about that particular piece of equipment through an editable database. Numerous software programs can be used for asset tracking and provide an open database platform for users to input and customize all sorts of information.

In BIM, the RFID tag uses XML to tie the external databases to the BIM file. Imagine being able to scan an RFID tag and pull up the information about a component while simultaneously having the software find it in a model. You could scan things such as doors, windows, hardware, HVAC equipment, furniture, and so on, and the information could be fed to the user in real time. Currently, there is no software that links to all the modeling software available, so this is pure speculation at this point. However, it doesn't seem too far of a stretch because companies such as Vela Systems have already begun using these tags to identify construction components and other components in the field.

Other options for linking the BIM to a database include directly inputting information into the elements' properties, which was demonstrated in Chapter 6. This can be done through the use of the software's default fields or the creation of custom fields that can show additional information about a component. Generally, this method is time consuming; however, it removes the need for linking a model component to an external database to drill down to information. In either scenario, the importance is to leverage the BIM to contain the needed information and remain a usable tool for the facility manager throughout the building's life cycle.

I believe that the order of BIM development is just now entering the facility management space in earnest. What began as a tool marketed for designers has transitioned into a tool leveraged by the construction industry and will eventually become a means of more effective operations. Many software vendors will say that their CMMS or facility management software uses BIM, and though that may be true for a portion of their application, there is a significant amount of improvement that can be delivered to this area for construction consumers.

BIM and Information Handover

The path to artifact and constant begins with data collection during design, continues through construction, and eventually transitions information over to the owner for use in operations in a meaningful way. As construction is wrapping up, there are often two overlooked factors that are critical pieces of construction delivery: commissioning and punch items.

Similar to other tasks that need to be completed toward the end of a project, less attention is given to items toward the end of the job, one of which can be commissioning. Commissioning for construction is defined as a process by the California Commissioning Collaborative as follows:

When a building is initially commissioned it undergoes an intensive quality assurance process that begins during design and continues through construction, occupancy, and operations. Commissioning ensures that the new building operates as the owner intended and that building staff are prepared to operate and maintain its systems and equipment.

Source: California Commissioning Collaborative

This process is an important aspect of quality control on a job. There is nothing more frustrating than finishing an installation of a system only to learn that it doesn't operate correctly. Because this represents significant risk on a project, certain controls, such as commissioning, are put in place to see that this doesn't happen. Typically, this process is manual and includes the use of spreadsheets and documents to monitor the installation and update testing of various pieces of equipment. The process is quite involved and includes performance and diagnostic tests for systems that are identified during the design phase as critical to the startup process (Figure 7.7).

Commissioning Process Overview

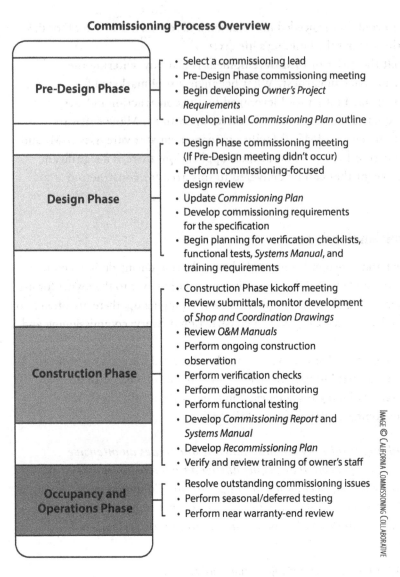

Pre-Design Phase
- Select a commissioning lead
- Pre-Design Phase commissioning meeting
- Begin developing *Owner's Project Requirements*
- Develop initial *Commissioning Plan* outline

Design Phase
- Design Phase commissioning meeting (If Pre-Design meeting didn't occur)
- Perform commissioning-focused design review
- Update *Commissioning Plan*
- Develop commissioning requirements for the specification
- Begin planning for verification checklists, functional tests, *Systems Manual*, and training requirements

Construction Phase
- Construction Phase kickoff meeting
- Review submittals, monitor development of *Shop and Coordination Drawings*
- Review *O&M Manuals*
- Perform ongoing construction observation
- Perform verification checks
- Perform diagnostic monitoring
- Perform functional testing
- Develop *Commissioning Report* and *Systems Manual*
- Develop *Recommissioning Plan*
- Verify and review training of owner's staff

Occupancy and Operations Phase
- Resolve outstanding commissioning issues
- Perform seasonal/deferred testing
- Perform near warranty-end review

IMAGE © CALIFORNIA COMMISSIONING COLLABORATIVE

Figure 7.7 Commissioning process

BIM has recently added significant value to commissioning agents because of its ability to monitor issue status, checklists, and progress toward systems installation. Additionally, there is the ability now built into tools such as Autodesk BIM 360 Field (Figure 7.8) to automate this manual process as well as use the visualization component of BIM to isolate systems or specifically tag model components that need attention. This ability means revisions issued are focused and can be tracked in real time. Ultimately, this revision process can be captured, or left integrated into a model such as Navisworks, which is tightly aligned to the artifact and constant strategy that tells the story of how a project is constructed through the use of static and model data.

Figure 7.8 Autodesk BIM 360 Field

Another aspect of importance as projects near completion is *punch lists*. Similar to the use case for commissioning, punch lists have traditionally been issued and completed in a low-tech manner such as printed plan sets, spreadsheets, or one-off applications to resolve issues in field related to quality issues of construction. Sometimes punch lists are managed by creating a room- or area-numbered spreadsheet in Microsoft Excel. The construction manager, architect, or designer responsible for punch list creation walks through the project, logs issues where they find them, and then makes a number of copies and distributes them to the subcontractors at progress meetings. When the subcontractor has fixed an issue, the construction manager marks the item as completed and hands the completed list over to the general contractor. When the architect completes the next walk-through, the designer lists the items as completed or not acceptable in the spreadsheet. However, as the architect is walking the site the second time, new issues might arise. Or if the project is a 30-story high-rise and the punch list is to be accomplished by going from floor to floor, then the amount of data may become overwhelming and the marks confusing. Specific locations to fix are cloudy especially if handwritten sketches and notes were the means of documenting the punch list.

Punch list items represent a unique opportunity for BIM use because of how much the process could be streamlined by embedding punch list information within the model from a single application such as BIM 360 Field, Latista Field, or Bentley Navigator (Figure 7.9). With the use of tablets, construction management personnel are now able to walk from room to room and provide detailed information about the punch list items and tag them in the model, issue them to the respective party for correction, and track status on resolution of open items.

Figure 7.9 Bentley Navigator

Some contractors prefer to use the composite model to organize and issue resolution items. Programs such as Autodesk BIM 360 Field and ConstructSim can be used to directly mark and comment on the BIM model (Figure 7.10). This method of coordinating punch list items is very effective and relies on the software's ability to automate the creation of sequential comments and view specific commenting. The ability to assign status to issues allows you to track project items to closeout. Additionally, the review tool tracks items to complete on the project through closeout. BIM 360 Field is particularly effective if you're already using it for in-field RFIs or quality control, because the team will have a basic grasp of the software and will be able to use the familiar interface to sort through the final issues of a project. Field does require some customization, as touched on in Chapter 6, so a BIM Manager or Autodesk representative should assist with the initial setup and creation of templates. Once templates have been created, they can be reused on every project.

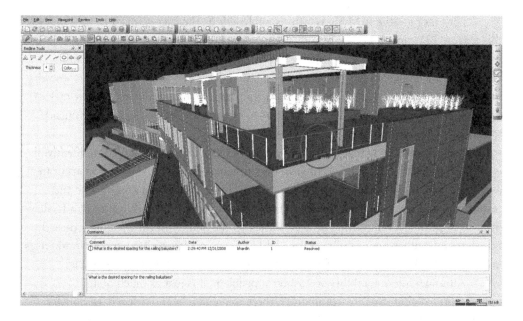

Figure 7.10 Example of Navisworks being used for punch list coordination

A disadvantage to using only the model for closeout is that in order to be commented on, linked to, or associated, an object must be modeled. This is another reason why the artifact and constant deliverable strategy works. Punch issue items such as "repaint this room" or "clean all stairwells of debris" can more easily be managed in a checklist, where BIM can be used to specifically identify where items are because they are located within the model, eliminating any confusion about location or object.

Since the last version of this book, closeout tools that leverage the model have come a long way. The use of mobile hardware has increased, and the ability to connect the dots between intelligent model objects and checklists via task-specific applications has resulted in significant efficiency increases for construction management professionals.

Maintaining the Model

Just as in the design and construction phases, the value of a BIM file to facility management is in its accuracy and ability to be up-to-date. Maintaining a BIM file with regard to facility management information is similar to maintaining the actual facility. As components are replaced, repaired, or removed, those changes will need to be reflected in the BIM file. The tool needs to be updated to match the facility so the accuracy doesn't fall behind and the BIM file becomes less of a resource for its user. If the BIM file is not updated and relevant, this makes finding the right information the next time more difficult. Without BIM, the facility manager refers to 2D drawings or self-inspects the facility. In the case of larger structures, this inefficiency will be clearly evident and reflected in the upkeep and potentially the stress level of the facility manager.

Let's face it—the BIM file is only as applicable as the user wants it to be. For this reason, BIM and CAD applications rely on the facility manager to update the files over the life cycle of the building. The exceptions to this rule are software updates, outdated data, change of personnel, archiving, and other tasks between the user and the information. Depending on the level of BIM training a facility manager has, the task of updating might require outside help in the form of software vendor support or outside technology consultants. Aside from these costs, however, the data for a facility rests squarely on the facility manager's shoulders. If a facility is to be expanded frequently, it is critical for the facility manager to keep the drawings and documentation as accurate as possible for future additions and renovations. This limits the amount of time and resources that a design team and contractor will need for their due diligence of a facility prior to beginning a remodeling project. Basically, the quality of the document directly translates into either savings or expense. And as firms migrate to BIM and as contractors use BIM technology, the value of BIM technology during a handoff will increase.

Although interoperability seems like a problem, it is just as much of an opportunity. Within the BIM community, there is a goal to standardize data transfer and develop software that is specifically geared toward BIM and facility management, removed from a CAD-based way of thinking. Although I anticipate that great strides will be made within the coming years in regard to facility management, BIM software, and computer-aided facility management (CAFM) technology, the process needs to be put in place to deliver complete native-format BIM files as well as all additional information.

The community should focus not only on how to apply a BIM database to room or spatial validation but also on how to use BIM for preventive maintenance, automated systems, component replacement information, and energy analysis, to name a few uses. Although the database of information is important, the facility manager faces the challenge of updating the BIM information with the changes made to the facility over the years. One of the biggest challenges in the facility management industry is how to use the data, manage it, and then change the data while maintaining the facility.

Current processes don't seem to be working as efficiently as possible, so how can we use new technology to change that? And is BIM the right solution? BIM as a software and as a process will be refined, and facility managers will ultimately benefit from it as a resource. However, there is still work to be done—which starts with the architect and ends with the facility manager—to document and keep the data current for the life of a building.

Ongoing Investment and Logistics for Facility Management BIM

BIM is gaining ground and is being more rapidly accepted and required by owners. BIM usage grew by 3 percent in 2003, 6 percent in 2005, and 11 percent in 2006 (according to the Eighth Annual FMI Owner Survey). In 2008, that number grew another 25 percent, and now 45 percent of BIM users say they are using it at moderate levels or higher and are seeing an ROI of 300 to 500 percent (source:

http://construction.ecnext.com/coms2/summary_0249-296182_ITM_analytics). However, according to the "Cost Analysis of Inadequate Interoperability in the U.S. Capital Facilities Industry" report, contractors and the AEC community at large aren't delivering linked and usable information to downstream users (owners and facility managers):

> *Owners and operators in particular were able to illustrate the challenges of information exchange and management due to their involvement in each phase of the facility life cycle. In summary, they view their interoperability costs during the O&M phase as a failure to manage activities upstream in the design and construction process. Poor communication and maintenance of as-built data, communications failures, inadequate standardization, and inadequate oversight during each life-cycle phase culminate in downstream costs.*

"Cost Analysis of Inadequate Interoperability in the U.S. Capital Facilities Industry," NIST 2004, by Michael P. Gallaher, Alan C. O'Connor, John L. Dettbarn Jr., and Linda T. Gilday. Available at http://fire.nist.gov/bfrlpubs/build04/PDF/b04022.pdf.

The answer to this question is in the amount of information construction managers compile and, more important, in the manner in which they compile it. It is essential to include warranty, cut sheets, and other data about a project, but the game is changing. Costs are rising. As a result, many owners are finding and requesting ways to save and be more competitive through the life of their investment. This will mean that contractors will need to change the information and process of delivering this information so that it can be used downstream to stay competitive. The BIM process during the creation of as-built documentation will change. The days of relying on the owner to interpret the mishmash of documents is over. Too many issues are at stake, including cost savings, energy conservation, and life safety, to name a few. The construction company that fully embraces BIM can realize the value of the technology and communicate that to the owner through education.

As discussed earlier, successful document and record BIM delivery involves having the operations personnel actively engaged from the beginning of the project. The operations personnel might very well have information about the CMMS that they intend to use to manage the facility, and how the documentation and model will need to interface with that software. In addition, the facilities group might have reasons for using the software they do because it might interface with accounting, human resources, purchasing, and other departments within an organization. For this reason, make sure that the way in which the project is being documented will work with the technology being used in the building.

Improved communication is one of the main thrusts behind the industry shift to BIM. However, using too many tools and too many formats with the intention of improving communication can be counterproductive to the group. When looking at a project holistically, consider the software that is to be used. Involving the operations personnel early on facilitates better input about the adoption of the model coordination plan and the information exchange plan and should help them define what level of detail and software is needed to perform their tasks.

Effectively documenting, operating, maintaining, and completing the life cycle of a facility are the ultimate goals of any facility manager. With a number of other tasks included within these descriptions, there is an enormous opportunity to better define these events. It's pure economics; if a building is designed and operated effectively through documentation, maintenance, and occupant satisfaction, then there is more profit to distribute. The building owner can then effectively use that capital to begin constructing other projects that keep the design and construction industry moving forward.

Many designers and contractors fail to realize that the ability for a building owner to develop new projects is directly related to how well a facility performs. Knowing this should directly influence all decisions for the downstream user. The architect should be maximizing efficiency, designing sustainably, and thinking ahead to the future of the building and working with the owner to not only define what the building is but what the building is to become 30, 40, and 50 years down the road. Likewise, the construction manager needs to be thinking about project documentation, operations, and maintenance manuals, as well as communicating and collaborating with the facility manager throughout the construction project to explain the intricacies of the facility before it is handed over. In turn, the facility manager must manage the BIM effectively and use it to operate the facility to its optimum performance while maintaining the health and safety of those occupying or working in the building and maintaining the life safety equipment such as sprinklers, alarms, and so on.

Training

BIM enables the facility manager to be more efficient; however, there are great strides to be made in the realm of BIM and facility management. With training and some experience, a facility manager can use the record BIM as a tool to more accurately manage a facility than ever before. Because CAD has been the technology predominately taught in universities and technical colleges, it has become the industry norm. For this reason, it is safe to assume that a facility manager who is handed a record BIM who doesn't have any prior training or experience with BIM or its respective pieces of software will have no idea what to do with it.

Therefore, the discussion of training needs to happen early on in the construction process for a number of reasons:

- First, by the time construction is completed, the facility manager will have a level of understanding to hit the ground running with the record BIM file to keep it current as the facility is handed over.
- Second, the facility manager is apt to give more valuable input into the design and end-level information available in the BIM file and will be able to inform the team what information will be most valuable at the end of a project.

- Last, the facility manager needs to receive the training for the next project. This is ultimately how we all have learned one way or another in BIM—through experience and use.

That said, the BIM-trained facility manager is a great resource for document accuracy and upkeep as well as for internally creating value for the building owner. Facility managers can extend their value to other associates without incurring any of the additional training costs for each associate. Although the decision might be to train all staff regardless, if a facility manager is in place who has been using it, that individual will be useful down the road when the newly trained associates have questions. Implementation planning is directly applicable to the facility management team, just as to an architecture or engineering firm.

Although there is still plenty of room for standardization in BIM—for example, the XML format exchange between all software systems or some version thereof—BIM somewhat negates the old discussions of lines, layers, and plot weights. Of course, the settings can be modified, but the fundamental difference in a standardized format between a CAD standard and a BIM standard in CAD is CAD's line-based process versus BIM's data-rich object-based process. The task of standardizing is somewhat of an arduous one, especially for an entire industry. Yet, the step that needs to be taken first is the step backward toward defining BIM's base language, then the industry can move forward.

Model Maintenance

To maintain a model, a facility manager needs to analyze what systems need to be put in place to be effective. For example, if the facility manager is managing a small strip mall, he or she might not need more than one license of the native software and probably won't require a sophisticated database to manage the facility. On the other hand, if that facility is a 1,200-bed urban hospital, then information systems need to be put in place to allow that facility manager to succeed. In this case, it might require multiple licenses of the native software, a server, sophisticated facility management software, and other infrastructure to support the operations of these facilities.

The implementation phase for a facility manager is no different in theory or practice from a construction company, architecture firm, or engineer's office. The general outline to support model maintenance is as follows:

1. Estimate the investment costs.
2. Develop an implementation plan.
3. Choose a manager or enable an existing manager to be BIM lead.
4. Support the user with training.
5. Support the manager with a team.
6. Learn more, contribute to the community, and develop processes.
7. Analyze implementation.
8. Monitor future trends.

Once the supporting plan has been developed, facility managers have some level of training, and the facility is in operation, a model edit workflow needs to be developed. As a best practice, it is usually a good idea to have one "model owner" who is responsible for updating and maintaining a model to reflect current conditions. Although this individual does not have to be the *only* staff member who is knowledgeable about a particular facility, this approach simplifies the job of updating, particularly in facilities with multiple sites that will have multiple model owners. Sometimes models are hosted and updated by third parties as a resource to assist as well.

Facility managers in general have a thorough understanding and knowledge of their processes, and they have in general used CAD drawings as supplemental information or have created their own. Ideally, in a BIM maintenance process the artifact information is updated and stays continually tied to model (constant) information over a project's life cycle. This is where BIM can create savings and efficiencies.

In the end, facility managers can be more efficient resources to the building owner and to the industry as a whole. Just as the development of BIM in the construction arena has continued to develop because of dialogue within the profession, introducing BIM into the facility manager arena and promoting discussion within this profession will advance both use and development of the tools.

One BIM = One Source of Information

Information in BIM is both visual and database driven. The concept of linking the visual representation with other information such as spreadsheets or other data sources is critical in operations. Within the BIM industry, many user experts refer to a model as "the single point of truth" or a project's "library." Either way, the concept is the same. Using a model that is the virtual representation of physical components, you can use the information beyond facilities and maintenance.

The future of operations will rely on software or web-based programs, such as ONUMA Planning System (OPS) and VEO by M-SIX—which are doing some great work in creating a platform that allows the freeform exchange and compilation of data to associate with models. These platforms enable models to "close the loop" and deliver useful data for future design and equipment purchasing decisions. Think about it. As BIM users, the construction industry is caught somewhat in a dilemma. There are silo tools that are useful to accomplish specific tasks for model workflows, such as Revit for model creation, Navigator for model viewing, Vico for estimating, and so on. Yet the underlying concept of unification of model data across isolated programs is what's been missing adoption for some time in our industry.

I have read countless articles and attended multiple presentations that claim the most important aspect of BIM is the "I." Many users take that to mean data integrity and quality, which is important, but at a high level there is a need for tools that support the use of systems information across multiple platforms. This concept extends beyond interoperability and is new to the construction industry, but it is critical. If you are to use different BIM tools and best leverage output from each of them, the use of platforms to host this data is going to become essential to future success. I often call this the "information backbone" of BIM (Figure 7.11). The information backbone is the repository of model data and other information that is aggregated and can be used for import into other systems or analyzed as stand-alone data points for Big Data analysis on trends and performance. This could then be applied a number of ways to make better decisions. For example, If you captured information in your facility management BIM tool that said you've had to replace a significant number of parts in a particular type of rooftop unit (RTU) over the last five years, you could make a value decision when designing the next project to look at previous performance indicators to see if you should select a different RTU.

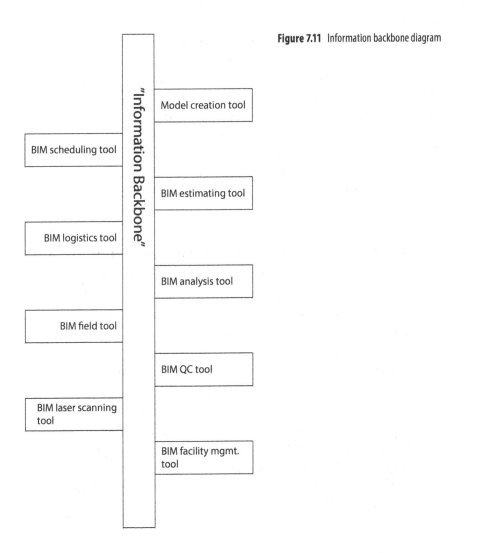

Figure 7.11 Information backbone diagram

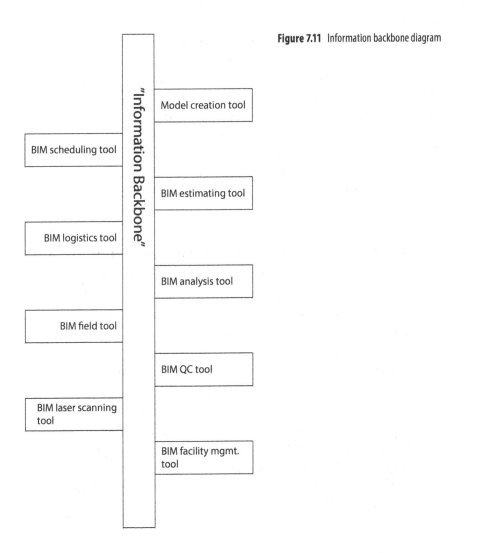

Another example might be the comparison of design BIM information for a mechanical cooling system that was to perform at certain levels, which you then mash up against an energy use or utility tool that says how the system is performing. This would then give you actionable information to see what is working and what isn't. Facility managers play a critical role in being the validating factor of design and construction assumptions to prove actual value.

Of course, this collection of data wouldn't be possible without the open and free exchange of information from proprietary software systems. The concept of open source programming or interoperability is part of the driving force behind the National Institute of Building Sciences (NIBS) and other organizations' efforts to allow software to talk with each other and for software companies to allow for customization, data extraction, plug-ins, and application development. This promotion of transfer of data is unique, and our industry has benefited from it. I'm not sure if we'll ever get to a point where all the tools that our industry uses are built on the same language, but as long as we can aggregate and sort the data from these systems we can prove the value. For example, websites are built using all kinds of tools that fit the needs or purpose of the user. Websites can range from simple plain-text formats to rich web content developed on different platforms such as Linux, HTML5, PHP, or even drag-and-drop design tools like Adobe Dreamweaver. What's unique about these languages is regardless of the code used to build them, if I open my Internet browser, it can read them all. Why is that? Of course, this is based on common web language and I believe that supporting efforts on interoperability are absolutely critical to future growth in BIM.

"If I Had Known That, I Would Have Written the Specs Better!"

Working on BIM projects often means that you learn about other team members just as much as you learn about your own profession. One of the first projects I had worked on as a BIM project was for a government owner group, and the contract specified that all design, construction, and project closeout deliverables were to be "in BIM." The project went smoothly, and both the design and engineering teams were very accommodating in model sharing. Overall, the project went very smoothly...until project closeout.

A month before project closeout, we realized for the first time that the owner wanted us to deliver a record BIM of the project. Aside from the formal construction documentation (RFI, ASD, ASI, and so on), we had done little to associate information to the model to meet the progressive BIM standard set forth by the owner—and we helped develop that standard! So, we set to work and began linking the specifications, PDFs, and other information into the BIM file.

> While working on producing the final product, we all came to realize just how poorly we had created the documentation that the operations manager was to receive. Although this was no different from the standard level of information that the owners had come to accept from other projects, we found what we were delivering needed bridges over serious information gaps—so much so that one of the architects responded to warranty information on an operable partition with this: "If I had known we were supposed to produce this at the end of the project, I would have written the specs better!"
>
> We learned from that experience that lots of information doesn't necessarily mean good information. It's the delivery of the right information that enables users downstream to succeed.

So why is all of this important? Simply put, our current model of design and construction assumes too much. Unless you are working under a Design Build Operate Maintain (DBOM) model, you are walking away from the answers to check against how you did. This has huge potential impacts on our industry. I encourage you to engage facility managers and owners in an ongoing manner to capture results and validate or disprove your assumptions. This can then be used by both the operator for better maintenance and the designer for better and validated decisions on design work.

Summary

Putting BIM as a resource in place for a facility manager decreases the amount of time it takes to get to and add information. BIM is still a tool that is only as valuable as the accuracy of the data input and the sophistication of the person using it. For this purpose, the process of managing facilities and what is to be expected from the construction and design teams at project closeout will change. Educated owners will demand more resources to equip their staff with the best tools to handle the job. These same owners have begun to hire from within the AEC community to better manage the information associated with their facilities, and contractors who aren't capable of producing this deliverable at the close of a project will be less competitive. BIM is effective as a single source of information through 3D representation and will be further refined and developed by what and how operations personnel use it throughout its evolution. Additionally, the best practice of artifact and constant deliverables increases effectiveness and information management efforts for facility operators.

Currently, a record BIM delivery is specific to each construction project. Although there are a number of guidelines for creating a record BIM (such as the

GSA's "BIM Guide for Spatial Program Validation" and "BIM Guide Overview" and documentation from the Open Standards Consortium for Real Estate), it has been my experience that the deliverable for each project will be different. This is not necessarily because of a lack of standards or interoperability but rather because the field of facility management and BIM are truly beginning to formulate just what one means to the other. Factors such as owner education, staff capabilities, and facility type all have a hand in defining what will be delivered at the end of a project and who will maintain it. The keys are to discover what is expected, educate the owner about BIM, create the record BIM to whatever level is desired, and implement strategies throughout the construction process to maintain it and extend its use.

The Future of BIM

This chapter takes a speculative look at the future of BIM: where it can go and where it is already headed. It will show what BIM will become in the coming years, as well as theorize on who will use it and how it will change the design and construction community.

What Will BIM Be?

BIM has become the catalyst for innovation in the construction industry. BIM has created a shift on two fronts:

- The first is the use of BIM for construction, where the use of models and related information creates opportunities for a better way of working.

- The second is the shift away from "business as usual" in construction. Discussions, hack-a-thons, presentations, and what-ifs are shining lights at older tools and processes with the core question of "Why do we do it this way?"

The resulting "technology renaissance" is a direct by-product of BIM's introduction into the design and construction landscape. Although BIM will continue to be a central enabler to better building, we will see tools and improvements that may or may not be BIM tools in the traditional sense. There is a need for apps and software that connect the model-related data to construction workflows. This area represents fertile ground for new tools and processes to improve or change construction as we know it.

Many of the predictions made about BIM in the previous edition of this book have come to fruition. Those predictions include:

BIM use will surpass 2D CAD use.   Because of the intelligent and parametric nature of the tool, CAD has lost ground against the new pace of design and construction.

Used effectively, a model will become a virtual replica of the physical construction to follow.   Software and hardware improvements continue to narrow the gap between the virtual representation and the physical manifestation.

Past processes and roles will continue to be blurred or disappear completely.   The real-world applications of BIM use will become a focus. Developments such as lower-cost laser scanners, photogrammetry, prefabrication, and software APIs have made BIM as a construction tool much more viable, as have advancements in true partnerships and collaboration that eliminates traditional roles.

An interesting dynamic to watch will be the introduction of real-time information and new technology into a builder's day-to-day tasks. What implications will Google Glass, wearable technology, and laser-scanning drones have in our industry? The opportunities are exciting, and those testing and proving use cases are perhaps the most compelling. I encourage you to test and think about ideas continually. Construction has made huge strides in BIM use and technology, but in many cases we are just now catching up to other industry peers. Accelerating our industry back to its former reputation as an innovative group of constructors is up to you.

Industry Trends

Our industry is now connected to seemingly infinite amounts of information that can be accessed in a matter of seconds, and if you don't make the move to select technology that supports its use, the industry will blow right by you. The value of intelligent parametric modeling is too great for owners, design-builders, or IPD teams to ignore when compared with the waste found in two-dimensional solutions. What is so promising about BIM is that it is not CAD, it is not drafting, and it is not lines on a paper representing the outline of a building. Rather, the model *is* the building.

Consolidated virtual design and construction models will continue to evolve to serve the needs of the team more effectively during design, construction, and beyond. As this book has outlined, current situations require a hybrid use of both 3D views and 2D drafting information to produce construction documents, and this will continue for some time. However, as BIM continues to grow in sophistication and experiences are shared, the use and detail of the models will change as well. The process of model development will continue to develop during preconstruction, carry into the field as an accurate construction tool, and then be used after occupancy in a much more organic life-cycle management tool.

The ability to create models that are construction-ready will steadily improve. In fact, questions as to why *representative* or *design intent* models are being created at all and if there's any value to them continue to persist. Currently, the ability to work in BIM requires multiple tools, and theoretically, the ability to have all team members continue to build on previously created models is not often possible. However, this book has outlined some strategies in the interim to promote this model evolution. Companies like Autodesk have developed the ability to host models in the cloud to further develop model-sharing strategies. Some companies are using server-based and virtual desktop environment (VDE) solutions that offer single model working strategies. As a result, the ability to build a structure based completely on parametric information will become a reality, just as models and the automation of construction will become a reality for more building components and potentially entire buildings soon. Scan the QR code in Figure 8.1 to see how close we are to unlocking that potential.

Figure 8.1 "3D Printers Print Ten Houses in 24 Hours"
(https://www.youtube.com/watch?v=SObzNdyRTBs)

BIM and Prefabrication

It isn't much of a stretch to believe that building technology will continue its evolution to use CNC and automated fabrication. Complex, custom structures require a high degree of accuracy and those building simpler structures are seeking higher profitability and better coordination. Both types of builders have the opportunity to reap the benefits of BIM as a tool for construction and prefabrication. In the future, many construction managers may be "assemblers," responsible for fitting the pieces of a building together as if it were a 3D puzzle on a jobsite to increase proficiencies and profit. However, the future will still require adept and technology-proficient professionals to carry out this work.

Figure 8.2 shows walls being fabricated directly out of the Revit model using two extensions: StrucSoft Solutions Metal Wood Framer (MWF) in combination with Weinmann's CNC program. Although two extensions were required to communicate with the Weinmann equipment, only one medium (Revit) was used to construct the model. What is truly amazing is the level of detail, accuracy, and sophistication that is capable. Scan the QR code (Figure 8.3) or watch the YouTube video produced by American Building Innovation of the entire process.

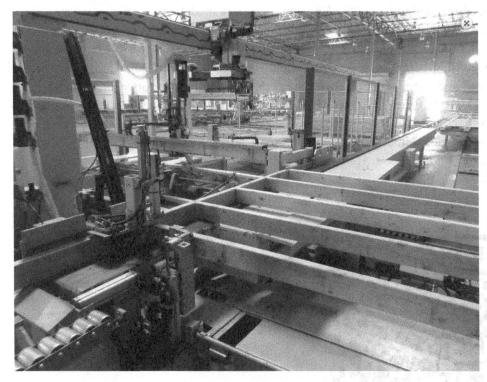

Figure 8.2 Fabricating walls in Revit

Figure 8.3 American Building Innovation video
(https://www.youtube.com/watch?v=VDWr2R_WKrQ)

Of course, nothing is perfect, and even if a team created the "perfect" model, issues would still arise throughout the construction of a building. The aim of BIM is to reduce the occurrence of these issues. Having process (BIM execution plan) *and* information exchange plans to facilitate increased communication between team members is critical to the success of a project. Our perspective on the industry overall (and in many ways the purpose for writing this book) is that the industry needs to realize that the old processes aren't working in today's world and will most certainly not work in the future. Even if old processes are working to some extent, they often aren't working as efficiently as they could be. Having seen the results of effectively using BIM from both the architecture and construction sides, it's difficult to not become staunch advocates for advancing the use and development of BIM and technology in construction.

BIM is here. Construction companies around the world are using it to sort through complex issues and increase the abilities of the design team. There are no guarantees that using BIM will give a team zero RFIs and the project will be finished without error ahead of time and under budget every time, but BIM has shown that it gets teams closer. As new processes, delivery methods, and teaming agreements are refined beyond even those mentioned in this book, the opportunities will be incredible.

New Processes and Roles

New processes are appealing to stakeholders in the design and construction profession. From the contractor's perspective, BIM represents an opportunity for reduced risk. As a result, contractor "contingencies" could be reduced, insurance rates decreased, and structures constructed more efficiently with better quality. Using the "build as modeled" philosophy, the contractor can have a much steadier foothold on otherwise shaky ground.

Additionally, architects have the opportunity to capture more relevance by moving away from design intent to more detailed and accurate models as well as becoming "information architects." There are many perspectives on the responsibilities and perceived liabilities associated with the architect and contractor collaboratively creating a construction-ready model; here are the two main concepts:

Contractor's Responsibility   The first perspective is that design professionals should continue by using the same processes as they always have and let the liability of document interpretation fall on the contractor's shoulders. Many of the statistics in this book show that this process lacks document quality and increases cost.

Architect/Designers' Responsibility   The second perspective is that there is serious risk in the profession of architecture in continually forcing more responsibility on the contractor by not accepting any. Some believe that the profession of architecture and engineering might face some level of extinction as a result of their unwillingness to take ownership of their designs by avoiding all risk and leaving it up to subcontractors, fabricators, and contractors to build their designs. This concern has some merit evident in the continued hiring of architects and engineers to become part of virtual construction teams working for general contractors. The popularity of one-stop shop entities (such as Jacobs, Parsons Brinckerhoff, and CH2M HILL) that offer in-house design and construction BIM services makes sense to many owners as the process becomes streamlined and the responsibility for the project becomes a team initiative, which further blurs the lines between professions.

Note: Not only have more BIM professionals been hired by general contractors, but specialty trades such as sheet metal, mechanical, electrical, piping, and structural steel fabrication shops are employing BIM specialists to meet the rising BIM demands.

The construction industry has begun to accept that new teams and responsibilities need to be created in order to realize the full potential of BIM. There are now more opportunities that influence how a project's fee is structured. Environmental analysis, code testing, and material use can be reviewed more quickly than before. Although these services can make for a better project, they should be carefully considered and outlined, because a team can become inundated with analysis and lose efficiencies on a project. On the other hand, innovative teams with new offerings can be compensated for new services that bring value to the owner and the project.

In the future, more teams will become aware of the wide range of benefits that come with partnerships. BIM has already changed the face of the construction industry. There wasn't as much talk of enhanced capabilities, reduction of errors, and future advancements with CAD with nearly as much vigor as we are seeing with BIM. Simply put, the enthusiasm has now been usurped by the reality and details of making the BIM process work. Although this is in no way an overnight effort, improvements are being made in both advancements in software as well as process changes and new concepts of delivery and teaming that strategically alter the way we design and construct buildings.

Interoperability

Software vendors and developers have listened to the design and construction industry's request for BIM to play better with other tools. Previously, interoperability issues required many pieces of software and many workarounds that took additional time and for which updates might or might not be available. Organizations like OPEN BIM Network, the National Building Specification (NBS), and buildingSMART International have pushed positive change working in this area and with software companies interested in making their products more usable by everyone.

The future of our industry will continue toward a more integrated course, yet the outcome of this integration is still to be determined:

- Some believe that general contractors will continue to merge and acquire design firms with the intention of streamlining practices and increasing profits.
- Others believe that project alliances will continue to gain market share and increase in popularity among the design and construction industries.

The bottom line is that our industry will continue to change traditional roles and processes.

As a result of this change, systems, enterprise resource planning (ERP) workflows, and transfer rates will need to work between programs. Organizations such as buildingSMART champion these critical efforts. In the future we will see more success stories with open standards, such as the buildingSMART IFC schema. For example, the Royal Institute of British Architects (RIBA) NBS launched the National BIM Library in 2012 that's based on these open standards. All models uploaded to the library follow their *NBS BIM Object Standard*, which states that it is "intended for construction professionals, manufacturers and other BIM content developers to assist in the creation of BIM objects that operate in a Common Data Environment (CDE)." By having a common language, these objects can then be leveraged for automatic specification generation out of the model, multilingual translation, consistent analysis, interoperability between software programs, and life-cycle management.

Purchasing decisions will be based on how well technologies communicate between other BIM systems. We should see a consolidation in the relative number of software tools available as features are integrated into new and/or existing software platforms. The barrier of software access will be reduced by vendors making them available on an as-needed basis via the cloud. Additionally, the disparity between how systems work together will be refined. The costs associated with the purchase of multiple pieces of software related to construction management will become more flexible, with options such as credits, pay-to-play, virtual desktops, and network licensing in the cloud.

The Future of Virtual Walk-Throughs

In 2009 I visited the Max Planck Institute in Tübingen, Germany to learn about their research in virtual reality. The institute sits high on a hill overlooking the humble town that's beautifully preserved with cobblestone streets and half-timbered houses. The simplicity of the town is quite deceiving as some of the most complex and advanced research in the world is conducted at Max Planck and since 1948 the society has had 18 Nobel Prize winners (www.mpg.de/183285/prizes).

Just a year prior to my visit, the institute, in partnership with the Technical University in Munich, the Sapienza University of Rome, and the Swiss Federal Institute of Technology in Zürich, created one of the most sophisticated treadmills in the world to study psychophysics. This was no ordinary treadmill; it weighed about 11 tons and cost over $3 million to create. When I visited, it was one of only two omnidirectional treadmills in the world and allowed researchers to study brain function by creating artificial stimuli with virtual reality, like making a subject walk a plank 100 feet in the air to create a false sense of fear and adrenaline. It made a lot of sense when I was told the other treadmill belonged to the U.S. government and was being used for military training exercises.

The treadmill, known as the Cyberwalk, was housed in a large room. It was positioned under a stagelike platform that had an approximate 12′ × 12′ opening in the floor exposing the treadmill walking surface underneath. You had to put on three items to interact with the Cyberwalk (see the following graphic): a harness, a helmet, and a virtual reality interface. The harness was for safety, the helmet was used to communicate your location in the room using sensors and cameras, and the interface was used to create the virtual world.

As you walked, the 11 ton machine would rotate beneath your feet, keeping you centered in the 12′ × 12′ surface. It was incredible how smooth this massive machine moved and how it reacted to your change in direction. (Scan the QR code in the following graphic, or visit `https://www.youtube.com/watch?v=oRK9IeCfYfE`, to see a video of me [Dave] walking on the Cyberwalk.) It was amazing! What-if scenarios flooded my mind. What if you could walk owners through their buildings before they're built? What if you could have construction teams literally walking around virtually installing their work, instead of sitting at desks all day? What if owners used this for raising awareness of their project and soliciting donations? What if superintendents could virtually walk sites and do interactive logistics plans instead of marking up PDFs?

I left Germany on a mission to find a way to incorporate BIM into what I had experienced. I knew it wouldn't take long for the industry to start exploring virtual reality and BIM. There were already "BIM Caves" and photorealistic walk-throughs being used, but the majority of them were clunky and left owners dizzy and looking for trashcans.

It wasn't until 2012 that I stumbled across a Kickstarter campaign that made me realize we were getting close. An 18-year-old named Palmer Luckey had created a virtual reality interface called the Rift. After reading about its promise of true immersion, I was sold on it being a key piece of the puzzle, but apparently I wasn't the only one. The campaign had over 9,000 backers and raised over $2 million. Two years later, in 2014, the Oculus Rift virtual reality headset was sold to Facebook founder and CEO Mark Zuckerberg for over $2 billion. Palmer Luckey no longer had to worry about the R&D budget.

However, the largest piece to creating true immersion, like what I experienced at Max Planck, was still out there. There had to be a way to walk around while using the Oculus Rift, and I knew that purchasing a $3 million treadmill wasn't the solution. It wasn't long

Continues

348

The Future of Virtual Walk-Throughs *(Continued)*

after I found the Oculus Rift that I discovered another Kickstarter campaign by Virtuix that claimed to have created a personal and affordable omnidirectional treadmill. It was cleverly named the "Omni." The campaign, similar to the Rift, exceeded its goal and raised over $1 million. It is the first of its kind, as shown in the following graphic. (Scan the QR code to see how the Virtuix Omni works.)

IMAGE COURTESY OF VIRTUIX

The Virtuix Omni is projected to weigh 75 pounds with a price point at $500 (according to their website, www.virtuix.com), which is *slightly* more mobile and affordable than the 11 ton $3 million big brother. It's truly amazing that my journey took less than 5 years, and a solution has been created that makes the state of the art "Cyberwalk" look archaic. The Oculus Rift and the Omni exemplify the speed of technology and the growing trend for faster, smaller, and better. These technological breakthroughs open the doors to endless possibilities with BIM and visual interaction. It may seem a bit like sci-fi, but we're literally at the point of interoperability where we could fly a drone around, laser scanning the cobblestone streets of Tübingen, upload that model into the Oculus Rift, and walk through the picturesque streets using the Virtuix Omni. If you apply that same scenario to a construction project, it gives a completely new meaning to a jobsite walk. This technology will allow true owner walk-throughs of the building before it's built. Potential donors can experience the owner's vision. It will allow detailers to virtually build like they used to in the field, which will reduce the risk of heart disease and diabetes caused by sedentary jobs. It takes BIM to the next level and brings even more clarity to the design and construction process. It is exciting to think of the possibilities that this technology creates—and this is just the beginning.

BIM and Education

For many construction professionals, BIM training begins in academia. Education, particularly in universities, is the perfect environment to challenge industry norms, experiment with new media, and explore what's possible. Schools teaching architecture, engineering, and construction management have an enormous responsibility and challenge to present their student body with technologies and processes that are relevant to the environment they will be exposed to upon graduation. These institutions must realize that teaching critical thinking and encouraging creativity is just as important as the software and technologies in current BIM coursework. Students make decisions on where to attend based largely on the following factors:

- A college or institution is often noted for what level of technological advances the school has embraced. If the university is deaf to the industry and its trends, this will be carried through in the school's reputation as well as its graduates. For this reason, it is critical that university systems be grounded in current issues and technologies, because these will be the issues their graduates will face in the industry.

- Institutions have the ability to be extremely advanced. There is little to no risk in exposing students to new technologies. Exploring new software and different processes should be done at this level, because it pushes students to learn and formulate views and solutions of their own. This will enable them to hit the ground running after graduation. This education might not result in a complete understanding of all BIM tools in the architecture or construction profession, but it should include a base-level knowledge of what the software is capable of doing and how it is to be used.

- Universities have the potential to save companies training costs. Although this shouldn't necessarily be a goal for universities since there are so many different pieces of BIM software, it is a marketable byproduct of embracing and training students on new software and technologies.

Fundamentally, BIM thrives in a collaborative environment and when it's used early in the design process. Schools need to create this environment of collaboration to better prepare their graduates for what to expect outside of their doors. This will enable students to better understand the working methodologies of construction teams and learn that few professionals work on their own. To some extent schools must test their students individually; however, creating team-focused groups with goals for success will prepare professionals for hands-on practice as well as the opportunity to explore groups that cross disciplines.

Universities have a unique opportunity to simulate real working environments by creating cross-functional teams of architects, engineers, and construction students. Each brings a differing perspective as well as a differing set of goals to the learning process. Understanding the importance each player brings to the team lays a foundation of collaboration as well as a frame of reference as to what to expect in the industry after graduation. Additionally, schools can use resources outside a construction department such as computer science, biotechnology, business and other departments to integrate and learn from other groups.

BIM software is complex, and developing an understanding of all the ins and outs of a particular tool often takes years. Additionally, the tools are constantly updated, thus creating a "moving target" in education. With this in mind, it is important to expose students to a basic level of understanding about how BIM tools work and what they can be used for. Encourage "deep dives" into tools if they're interested, but avoid the granular "picks and clicks" training. Although some students might continue to learn about other pieces of software after graduation, it's important for schools to teach BIM at a conceptual level.

Often students will begin learning whichever software is chosen to be part of the curriculum and will self-teach to a more advanced level. This is a unique characteristic of GenYs and millennials. This generation has the ability to learn software by simply working in it, clicking icons and working within the interface. This level of comfort with technology, such as using the Internet, playing video games, and leveraging the power of computers for advanced problem solving, is sure to continue as a trend for generations to come.

In essence, instructors cannot afford to believe that BIM tools are too complex and varied, and avoid spending time researching and understanding the software. It will be critical for instructors and schools alike to embrace some level of BIM technology and introduce students to these programs and concepts, which can add a new layer of knowledge to students' learning and increase their overall marketability.

Schools should investigate BIM and its uses within the construction industry holistically. BIM has a significant role in the new way of working with particular areas of focus such as design apps for architects, analysis and calculation features for engineers, and model management theory for construction management students. Many job opportunities now exist in design and construction that are centered on BIM and technology. Architects are being asked to use BIM to design sustainable buildings that capture rainwater, use less energy, and are more eco-friendly over the life cycle of the building.

Engineers are being asked to use BIM to calculate complex computational fluid dynamics (CFD) equations, find what type of steel is most efficient and least wasteful, and analyze information from a building to measure performance. Contractors are becoming information managers whose field of expertise extends beyond just construction management. Management of data transformed into information that can drive better estimates, schedules, performance, and connectivity will define the success of the contractor of the future. These are all real-world components of the modern AEC industry, and schools have a great opportunity to begin passing this information on to students.

BIM and the New Construction Manager

BIM's biggest opportunity in the future will be the direction that professionals find useful value. Sophisticated owners are seeking out design and construction companies that use BIM to meet their needs. Because educated owners will award projects based on qualifications, team selection will be critical to the success of these companies. These BIM-enabled companies rely on the professionals they hire who make up these teams to deliver results. Not only is choosing the right associates critical when beginning a BIM process, but these team members will be the ones to refine it over time. Wanting everything to be BIM enabled from the word *go* is not a reality and takes time to develop.

Our industry must continue to determine how BIM technology can increase profitability by finding associates who can produce process-based results using these various technologies. The need for measurable results does not necessarily have to be compared with CAD anymore; rather, these associates should begin comparing BIM-enabled projects with other BIM-enabled projects. The case for BIM has been made. And when direct comparisons are made, we will begin to see results that compare apples to apples in both process and technology refinement specific to BIM.

When direct analysis is accomplished, we'll see a much clearer picture of where we can be effective and where the technology needs to grow and in which direction. With the focus on developing and defining procedures, we'll get the power of people sharing information and experiences and lessons learned along a related timeline and defined milestones. In addition, entrepreneurs will develop the software, resources, and dialogue necessary to meet the industry's needs that are defined along the way. Where gaps exist, construction companies may develop their own tools to get them where they need to go. This ecosystem of innovators that create new and improved applications is critical to the success of BIM in our industry and is a driving force behind writing this book in this exciting time of education, research, and implementation of this new technology.

So, who are these people capable of using BIM technology, and where will they come from? Just as important, who currently is filling this industry need?

Currently the construction industry is hiring architects, engineers, and technology professionals who have received specialized training or have prior experience in BIM. The need for technology innovators will grow and the field will become more competitive. Though these resources may not be "BIM only" personnel, there is a growing trend to hire staff with a dedicated focus on technology and process improvement.

Future generations of people who use more advanced tools will be the ones who continue to develop BIM. These future architects, engineers, and construction managers will cross conventional industry lines and focus on successful project delivery. In fact, it is in their very nature to leverage technology to access information and work together as teams to accomplish tasks. This future generation expects ease of access to information: Over 97 percent of them own computers, and over half of them use blogs to disseminate information among their peers and colleagues. In the construction industry these new professionals are part of a larger interdependent relationship, in which they will require mentors to guide their intuitive and proficient use of the software tools by providing insight from real-world experiences and lessons learned.

These professionals will bring a level of understanding and comfort with software and technology that more senior members of the team might not have. Moreover, they will be focused on using all resources available to deliver projects. In this book, we have talked about the virtual construction manager, or BIM manager. The rapid expansion in this previously nonexistent field, where field engineers, project managers, and superintendents are now trained in BIM, allows BIM management to further improve processes and facilitate innovation.

BIM Manager Definition

Throughout the book you have heard the terms *BIM manager/director*, *virtual construction manager*, *VDC coordinator*, and so on. In many ways, this position is responsible for virtually constructing, analyzing, and managing a model. The responsibilities assigned to this role vary from company to company, yet the following are recurring themes in the job requirements for a BIM manager:

- Take responsibility for the entire scope and quality of the BIM team.

- Ensure adequate personnel are available for each BIM project's needs.
- Recruit qualified modeling personnel, and provide management assistance to the BIM recruiting efforts.
- Participate in selecting BIM projects to be pursued.
- Attend preproposal meetings, perform site visits, and collect data required to perform modeling tasks.
- Assist with responses for requests for proposals by being familiar with Industry standards and terminology as well as listing past project experience and examples.
- Prepare conceptual and detailed modeling budgets for proposed or awarded projects, and review them with the operations manager.
- Monitor and manage BIM expenses, and review them with senior management.
- Assume key roles in select sales presentations for assigned projects.
- Identify the modeling team for active or proposed projects.
- Ensure the responsibilities of others and assignments are coordinated and met.
- Guide model creation/assimilation throughout the preconstruction process.
- Provide assistance to the estimating team for model review and RFI model documentation.
- Take the lead role in facilitating, conducting, and participating in project kickoff meetings, design meetings, preconstruction meetings, and the operation's project meetings.
- Establish credibility and confidence for BIM as needed with clients, design team, subcontractors, and internal team members.
- Maintain and expand various design relationships and partnerships.
- Develop and ensure BIM policies and procedures are implemented and followed on projects.
- Provide guidance in regard to design management and model completion.
- Conduct detailed model reviews to ensure project expectations are met.
- Ensure questions regarding projects and model information are properly documented and addressed to the design team.
- Ensure measures are in place to monitor and track changes, and provide information to internal team members, design team, and subcontractors.
- Proactively identify risk factors with senior management.
- Assist operations with the development of construction phasing models throughout the project as necessary.

Continues

354

BIM Manager Definition *(Continued)*

- Assist operations with the development of project site utilization modeling and logistic coordination.

- Encourage the exploration of innovative, technically creative model presentations as needed.

- Strive to attain 100 percent customer satisfaction for BIM projects.

This role is relevant today, but as the traditional majority, as mentioned in Chapter 6, "BIM and Construction Administration," begins to embrace BIM, these roles will be absorbed into companies and project teams. BIM will be the *new* business as usual, and BIM-specific titles will start diminishing. Construction management firms will begin looking for chief technology officers and technology directors to act as liaisons between architects, engineers, contractors, and facility managers to better understand their needs, interdependencies, and ways to manage Big Data. Additionally, this entity will be tasked with understanding industry trends, keeping the company relevant with new technology, and identifying and resolving any deficiencies within the organization. They will become a critical piece of the overall practice and marketing strategy of the company.

BIM and the New Team

BIM encourages the formation of new teams in the industry. In the form of more integrated, project-driven teams, BIM will drive the industry to refine the way it works. BIM as a technology has the ability to transform our industry and the future of other industries, such as geographic information system (GIS), accounting, project management, emergency response, global positioning system (GPS), and environmental applications, among others. The fact that there are more intelligent models created means there is more information. When this additional information is more effectively shared and linked to other systems, additional savings, and reduction in duplicative work, will be realized.

As BIM grows, the following scenarios are becoming reality:

- As they are driving toward a building fire, firefighters can pull up the building's 3D model from the city's database to identify the location of fire extinguishers, shutoff valves, and emergency exits before ever stepping off the truck.

- Code review professionals can perform more accurate code reviews through the use of software (such as Solibri) to identify necessary clearances, heights, ratings, and appurtenances.

- Government agencies can map their GIS systems with 3D information to show zoning and adjacent heights, and use and simulate natural disaster management, attack, and energy failure, among others.
- Environmental agencies can simulate buildings' energy use, heat gain, and carbon footprint.
- Government organizations and armed forces can use BIM, radio-frequency identification (RFID), and 3D mapping to manage assets and personnel equipment and to simulate everything from temporary base setup to operation simulation.

Ultimately, BIM is a usable tool for many other stakeholders involved beyond the design and construction communities. BIM will become a better resource than previous technologies for opportunities to mock up and fabricate systems, link to other software, and remove unknowns prior to physical construction (see Figure 8.4). As BIM is still being defined in many ways in the construction industry, we can expect to see other groups find purposes for which BIM can be used to benefit them as a result of a single and current database of information.

IMAGE COURTESY OF McCOWNGORDON

Figure 8.4 Ronald McDonald House built using BIM and prefabricated mockups

Increased data flows will create an entire workforce dedicated to information management. This enhanced data control will come in the form of virtual design and construction teams. As newer staff members continue to rise within organizations around the world, the comfort of working with such technologies and the means of increasing the individual's effectiveness will create much leaner and more focused groups. The value of professionals using software to construct, test, simulate, and analyze prior to erection will increase and shift the norm. Likewise, the tools will increase in sophistication and usability as more tools become interoperable and as the industry defines its processes and informational needs. The new team will have the benefit of industry BIM experience and will be responsible for collectively delivering to the project's stakeholders and owner as joint teams with a clear understanding of the information needed, its uses, and its importance.

In many ways, this has already happened within the design and construction communities. Design-build firms currently leverage their ability to deliver projects faster and more efficiently because all of their operations are under one roof. The future of the BIM team will probably be one of collaborative model creation in real time, just as Microsoft Office 365, Google Docs, or Bluebeam Studio exist today for document creation. We're almost there, but in the future, technologies will be developed to eliminate the current shortcomings in BIM. More important, there will be new teams with new perspectives to further enable BIM technology developments, driven by a focus on information.

BIM and the New Process

Throughout this book we have talked about process. Creating a BIM process is often at the mercy of the current software available. At the same time changes made to the design and construction of a project must be profitable in order to be worthwhile. Process change, technology implementation, and the addition of personnel behaviors all take time to develop.

Shifting from CAD technology and CAD thinking is something that will be done over time, not overnight, and there are many hurdles to getting BIM integrated into a company. Implementation strategies can easily take years to fully realize, staff is difficult to find, and using BIM requires the use of new processes. Additionally, new technology and tools continue to enter the market and might further shift processes and thinking, perhaps even beyond BIM as we know it today. However, when compared to the potential results, these challenges make the case for BIM. At times, BIM can be daunting and somewhat overwhelming due to the new influx of information and tools. This does not mean that there is room to be laggard about beginning down the path to implementation and still remain competitive in today's construction market. BIM will gain market share and its users will be selected by educated owners and fellow professionals, and supported by facility managers.

When you define process in tandem with available software systems and milestone tasks, you create a roadmap for future developments as well as a tool to learn from. BIM is not the be-all, end-all solution to technological developments in the design and construction industries. If BIM is to succeed, it must serve as a conduit of information exchange by transforming the way we practice design and construction. In this regard, BIM has been an invaluable tool in opening the dialogue about new delivery methods and collaboration methods and in questioning our existing processes as an industry to find a better way to enable information flow.

Future Opportunities

BIM continues to grow with new applications and new abilities. The past decade has seen a rise in BIM applications, and although it might be hard to believe that more tools will enter the market, keep in mind that older tools will fall by the wayside as a result. For example, some of the tools that exist today are automated code-checking, clash detection, model printing, simulation, laser scanning, model-to-field layout, estimating, and sustainability analysis software—all developed within a relatively short time frame. So what comes next?

Our industry could see a completely digital means of delivering a construction project that uses no paper drawings and solely relies on a well-constructed BIM file for everything from "e-permitting" with local municipalities to contractors directly ordering and fabricating materials from downloaded BIM components.

There's a growing opportunity to leverage models for prefabrication, described throughout this book and shown in the American Building Innovation video (Figure 8.3). This is not only an opportunity, it's a necessity due to growing populations, depleting natural resources, fast-track delivery methods, shortage of skilled laborers entering the construction industry, and the safety of the labor force. BIM provides a solution to all of these issues through the use of prefabrication. Once you've watched the American Building Innovation video, scan the QR code in Figure 8.5 (or visit https://www.youtube.com/watch?v=YILAxkiYcxw) to watch a video produced by FPInnovations' Forintek division; then scan the QR code in Figure 8.6 (or visit https://www.youtube.com/watch?v=tJ735VaOIqY) to view the prototype animation of TheoBOT by Theometrics.

Figure 8.5 "Stick Built vs. Prefabricated Wall Panel House Construction" video

Figure 8.6 TheoBOT

There are also opportunities within the manufacturing market to begin creating parametric components that contain embedded specifications and life-cycle information. These components should automatically host to other components based on the applicable code and use. Future possibilities might be SMART exterior skin systems that automatically generate ideal waterproofing details based on the adjacent material's product data or windows with embedded specification information that notifies the designers when they don't meet the natural daylighting levels. This sort of BIM automation creates a new level of design and accuracy. Although implementation will require an investment on the behalf of manufacturers, many manufacturers will undoubtedly see this as a temporary opportunity to gain an edge over their competition (until they create their own). Additionally, manufacturers will be able to limit the amount of paper used for brochure and catalog printing, establish themselves and their commitment to the design community, and provide a resource that further promotes their products. Third-party websites that link to all of these manufacturers, like the National BIM Library, will rise in popularity as well.

In the future, we will see more software combining multiple analyses and testing tools into a single product, like VEO by M-SIX. This will in turn expedite the efforts being put forth by the design and construction teams. Multiple analysis test beds (MATs) that can test for constructability, model integrity, and clash detection at the same time will be needed, in part because of the availability of other tools entering the market and the need to streamline the process of testing the BIM thoroughly prior to construction. MATs will be able to test a model simultaneously for daylight and heat gain, system performance during peak needs, and CFD of the building design.

Because of the number and cost of BIM tools, combining these resources into a single platform will be somewhat of a necessity. In the future we will be able to send a model via the web to a series of linked MATs and return after the model has been tested by a large array of tools. Although companies like Assemble and Sefaira currently offer this service, many other companies and testing software will follow suit and adopt an online test bed strategy that design teams can plug into. Ultimately, this will better enable the designer and builder to make better decisions on many fronts.

Future Relationships

In his book *How Buildings Learn: What Happens After They're Built* (Penguin, 1995), Stewart Brand describes how structures evolve and are refined and altered over time, arguably creating better architecture and purpose. What's interesting is how a design that was originally intended for a singular use evolves. For example, the factory became a warehouse that then became artists' lofts, which then became condominiums and first floor storefronts. Structures can be quite complex, ranging from hospitals to factories to laboratories, and change over time. Additionally, owners' needs, construction materials, construction technology, and available resources are constantly changing as well.

Constructing and remodeling facilities requires an advanced understanding of codes, construction methods, safety, materials, design, use, construction scheduling, and holistic life-cycle thinking. Although it is possible, it is highly improbable that an architect will have the required construction and field experience to successfully construct a building and coordinate equipment, subcontractors, vendors, inspections, and safety protocols, all while maintaining the project schedule. It is just as unlikely that a contractor will have an understanding of codes, design logic, programming, accessibility issues, material properties, fire assemblies, and so on, to successfully design and document a facility prior to construction. Thus, we will need collaboration now and in the future of our industry. However, in the future we will see an increased need for flexible teams who continue to blur the lines of common practice and push the boundaries of integration and use of technology automation and augmentation.

By combining the knowledge base of all parties, you can begin to create a recipe for successful project delivery. BIM is a platform that enhances that collaboration and allows groups to learn from one another as the project is being developed and to be more quickly informed by useful project information. Flux, Google's first company to develop out of the moonshot labs, embraced this concept of open source intelligence to create a data-driven design platform. The idea is very similar to what Sean Gourley, co-founder of Quid, calls "augmented intelligence." It's letting the computers do the heavy lifting so that your team can focus on bigger strategies and innovations. It optimizes the design process based on the planetary, human, and building relationships.

When you think about it, there's accessible planetary data available to us for sunlight, wind, soils, vegetation, climate, and rainfall. There is human data of capital, comfort, and use of the spaces. Lastly, there is building data-like types, available footprint, adjacent buildings, materials, entitlements, zoning, and codes. The key is finding the interrelationships of the data in order to automate design options and make quicker and smarter decisions as a team using collective knowledge. The human mind cannot process all of the interrelationships fast enough, and we spend more time manually sorting through information, which means we rarely have time for innovation. This new process of team relationships augmented by computer processing is an intriguing idea and could lead to the most innovative ideas we've ever seen. As Erin McConahey, engineer and team member of Flux, so eloquently put it in a presentation at GreenBuild 2014, "It's just math."

The future of BIM is both exciting and challenging. Current opportunities are allowing us to bridge the gaps in a disconnected process more and more. This begins in new education standards and carries through to new ways of managing projects, mentoring, marketing, ordering, managing, and quantifying results. It has been our experience in working with BIM that there are many more opportunities than shortcomings and many more possibilities than liabilities. In construction, very little is constant and things are always changing.

Ultimately, the flexibility this industry requires needs to be reflected in the technology and processes used by its professionals. Additionally, the professionals in the industry are critical to its success. The behaviors and the experiences gained and shared are critical to the growth of a BIM-enabled process. Industry professionals who have a working knowledge of the needed tools and problems to be solved will continue to explore the necessary gaps and new methods that should continue to develop the software required. BIM has proved to be an extremely valuable tool to many in construction and will be one of the most exciting developments in the years to come.

Virtual Builder Certification

As you've read through this book, you've probably noticed a common theme to BIM's adoption, success, and innovation; it's all about the people. Our world is faced with environmental, information, and human challenges. Preventing global warming and preparing for exponential population growth falls on the shoulders of this generation. If you've read this far, then you must share the same passion and belief that BIM can drastically improve our efforts in these issues. We believe that BIM has the ability and power not only to transform the construction industry, but also to improve the quality of the built environment. So how does the use of BIM translate to an improved world?

- Structures that are designed to save energy, constructed with a high degree of quality, reduce our environmental impact.

- More information translates to better decisions, whether that is material choices, safety options, construction type, or anticipated operational costs. Everyone benefits from better, more useful information.

- Innovation is a journey, not a destination. BIM provides a means for us to understand and improve the way we build. In the future, visionary owners will push creative teams to reinvent our industry.

We all want to make a difference in the world. Virtual Builders was established to help realize those opportunities in an open source format to promote growth and innovation. Its goal is to break down the silos of owner, architect, engineer, and contractor and create a team of people and companies committed to openness, awareness, enthusiasm, and consistency. It is a group dedicated to making an impact with BIM and technology in the design and construction industry.

Openness   Not organization or vendor specific like AIA, AGC, LCI, DBIA, Autodesk, Bentley, Trimble, or USGBC, but advocates the collective beliefs of all of them and supports an open source means of sharing information

Awareness   Educating on best practices for processes enabled by technology

Enthusiasm   Passionate about technology, continuous improvement, and celebrating the successes of its members

Consistency   Giving the industry a standard of performance for delivering BIM services on a project

BIM touches many aspects of what we do in this industry. One way we can start working together to build better-quality, sustainable, and high-performance buildings is through organizations like Virtual Builders. We encourage you to get involved; to receive credentialing through Virtual Builders simply means that you are committed to this goal. For more information, visit the website: www.VirtualBuilders.com (Figure 8.7).

COURTESY OF VIRTUAL BUILDERS

Figure 8.7 Virtual Builders

Summary

The construction industry is going through a remodel, and we're currently in the middle of the most radical phase our generation has ever experienced. The walls of titles and backgrounds are slowly disappearing and new spaces of collaboration, information sharing, and innovation are being created. It's no longer about being a project manager in a corner office—it's about being a part of a networked and integrated team. The intellectual power once held by the individual is now at the fingertips of the fastest Googler. The ones who will remain after the remodel are the ones who understand good *process*, who are empowered with *technology*, and who embrace the *behaviors* of this new culture.

Index

Note to the Reader: Throughout this index boldfaced page numbers indicate primary discussions of a topic. *Italicized* page numbers indicate illustrations.

NUMBERS

3D BIM, 15, **50–52**
3D printers, 30–31, 116
4D BIM. *see* model-based scheduling
5D BIM. *see* model-based estimating

A

accuracy, installation, 234
activity tracking, construction, **234**, *235*
addenda
    BIM. *see* BIM addenda
    definition of, **65**
AE (architectural and engineering) models, 52, *55*
AGC (Associated General Contractors of America), *64*, 65
AIA (American Institute of Architects), *64*, 65, 66
Alberti, Leon Battista, 46
analysis
    building codes for, **179**, *179*
    building rating systems for, **177–178**, *178*
    concrete CO_2 emissions, **179–180**
    data, **27–29**
    model-based, 74–75
    multiple, 358
    Sefaira for, *182–187*, **182–188**
    software for, 175–176
    sustainability, **180–181**, *181*
animation, scheduling, *221–226*, **221–226**
Apple Watch, 22
AR (augmented reality) simulations, **115**, *115*
architect-controlled record models, 264
architects
    in ConsensusDocs 301, 65
    DB delivery method and, 59–60
    new responsibilities of, 344
    uses of BIM by, 351
architectural and engineering (AE) models, 52, *55*
Architecture 2030, **176–177**

artifact deliverables
    CAD files, **314–315**
    constant deliverables and, 315–316
    hybrid approach to, *316*, 317
    overview of, **310–311**, *311*
    PDFs, **311–312**, *313*
As-builts—Problems & Proposed Solutions (Pettee), 310
Assemble Systems, intuition and, **286–287**, *287–288*
Assemble tool, for cost trending, **172–175**, *173–175*
Associated General Contractors of America (AGC), *64*, 65
attention span statistics, *41*
augmented intelligence, 359
augmented reality (AR) simulations, **115**, *115*
Autodesk BIM 360 Field
    barcodes/QR codes in, **297–298**, *298*
    commissioning in, 326, *327*
    equipment database in, *301*
    features of, **291**
    mapping equipment to, **291–295**, *292–295*
    mobile application, *297*
    to status material, **299–301**, *300*
    uploading information into, **295–297**, *296*
    visualizing equipment in, **301–303**, *302–303*
Autodesk BIM 360 Glue
    email invitation, *159*
    real-time clash alert, 27, *28*
    sharing models, 291–292, *292*
    uploading models to, **159–160**, *160–163*
Autodesk Communication Specification, 75–77, 79, 83
Autodesk Navisworks
    clash detection in, *205–207*, **205–208**
    Comments tool in, **243–246**, *244–246*
    default units in, 219
    features of, 198
    field information via, 242–243
    importing search sets into, **288–290**, *288–290*
    NWD/NWF file formats, 198, 217, 219

W